MONTANA ROCKS!

A Guide to Geologic Sites under the Big Sky

Robert C. Thomas

Illustrated by Chelsea M. Feeney

The Geological Society of America
Boulder, Colorado
2026

GEOLOGY ROCKS!

A state-by-state series that introduces readers to some of the most compelling and accessible geologic sites in each state.

First Printing, May 2026

Printed in the United States by Versa Press, Inc.

All photos by Robert C. Thomas unless otherwise credited.

Cover photo: Mt. Gould in Glacier National Park. The prominent, dark layer is an igneous intrusion that baked the enclosing limestone into white marble.

Library of Congress Cataloging-in-Publication Data

Names: Thomas, Robert C. (Robert Curtiss), 1962- author | Feeney, Chelsea McRaven, 1980- illustrator
Title: Montana rocks! : a guide to geologic sites under the big sky / Robert C. Thomas ; illustrated by Chelsea M. Feeney.
Description: Boulder, Colorado : The Geological Society of America, 2026. | Series: Geology rocks! | Includes bibliographical references and index. | Summary: "This book tells the geologic story of Montana over the last four billion years. Sixty sites were chosen based on quality, accessibility, and beauty, and they are laid out in chronological order from the origin of our continental crust to the human experience with our geologic heritage. Read the book to learn the geologic history of Montana or use it as a travel guide"— Provided by publisher.
Identifiers: LCCN 2026003314 | ISBN 9780813741277 paperback
Subjects: LCSH: Geology—Montana | Geological time—Montana | Montana | BISAC: SCIENCE / Earth Sciences / Geology | NATURE / Rocks & Minerals | LCGFT: Guidebooks
Classification: LCC QE133 .T46 2026
LC record available at https://lccn.loc.gov/2026003314

PUBLISHED BY:
The Geological Society of America
3300 Penrose Place | P.O. Box 9140 | Boulder, CO 80301-9140, USA
+1.303.357.1000, option 3 | +1.800.472.1988
gsaservice@geosociety.org | www.geosociety.org

DISTRIBUTED BY:
Mountain Press Publishing Company
P.O. Box 2399, Missoula, MT 59806 | +1.406.728.1900
+1.800.234.5308 | info@mtnpress.com | www.mountain-press.com

Preface and Acknowledgments

This book tells the geologic story of Montana over the last 4 billion years. The sites described in this book were chosen based on their quality, accessibility, and beauty and are laid out in chronological order from the origin of our continental crust to human interactions with geology. You can read the book at home and learn the geologic history of Montana, or use it as a guide as you travel the state and immerse yourself in the deep-time story. Site maps are included to help you find each geologic marvel, so all that is required is your time and curiosity.

Our understanding of Montana's geology was made possible by countless scientists who spent lifetimes climbing mountains and weathering storms to gather geological data. They analyzed lab results and tested ideas until the best explanations emerged. Late nights were invested in putting ideas on paper that were then launched into the brutal universe of peer review to be scrutinized by contemporaries with appropriate credentials. Good ideas survived, and bad ones were discarded, and we rejoiced when we were wrong because that is when we learned. Evaluating and testing ideas is the beating heart of the scientific method, and this book is firmly rooted in this painstaking process.

Many people helped to improve this book. Discussions with Julie Baldwin, Rod Benson, Bruce Bjornstad, Darrel Cowan, Colleen Elliot, Dave Foster, Bill Fritz, Chris Gammons, Yann Gavilott, Emily Geraghty-Ward, Richard Gibson, Rick Graetz, Debbie Hanneman, Jack Horner, Don Hyndman, Tom Kalakay, Jeff Kuhn, Paul Link, Bekah Levine, Bill Locke, Jeff Lonn, Steven Losh, Marli Miller, Dave Mogk, Lisa Morgan-Morzel, Jesse Mosolf, Stuart Parker, Anneliese Ripley, Sheila Roberts, Spruce Shoenemann, Jim Sears, Derek Sjostrom, Larry Smith, Mike Stickney, Jeff Tepper, Susan Vuke, Don Winston, and Nick Zentner were invaluable. They are not, however, responsible for any errors. Tyler and Katie Steele with Shade Tree Outfitters in Fort Smith provided me with transportation to "the monocline" in Bighorn Lake. Some photographs and artwork were graciously provided by Rich Aram, David Bennett, Ron Blakey, Eric Eliasson, Richard Gibson, Doug Henderson, John Lambing, Dustin Leftridge, Steven Losh, Mike MacLeod, Travis Mahn, Montana Historical Society, Chris Pace, Jen Pierce, John Sanford, David Varricchio, and Scott Williams. All sources of illustrations modified for use in the book are credited in the figure captions. Chelsea McRaven Feeney's illustrations made the words understandable, and the organizational and editing expertise of Jenn Carey turned it all into a book. My deep appreciation goes to Anneliese Ripley, Abbey Thomas, and Haley Thomas because time lost with family is an inevitable consequence of writing a book.

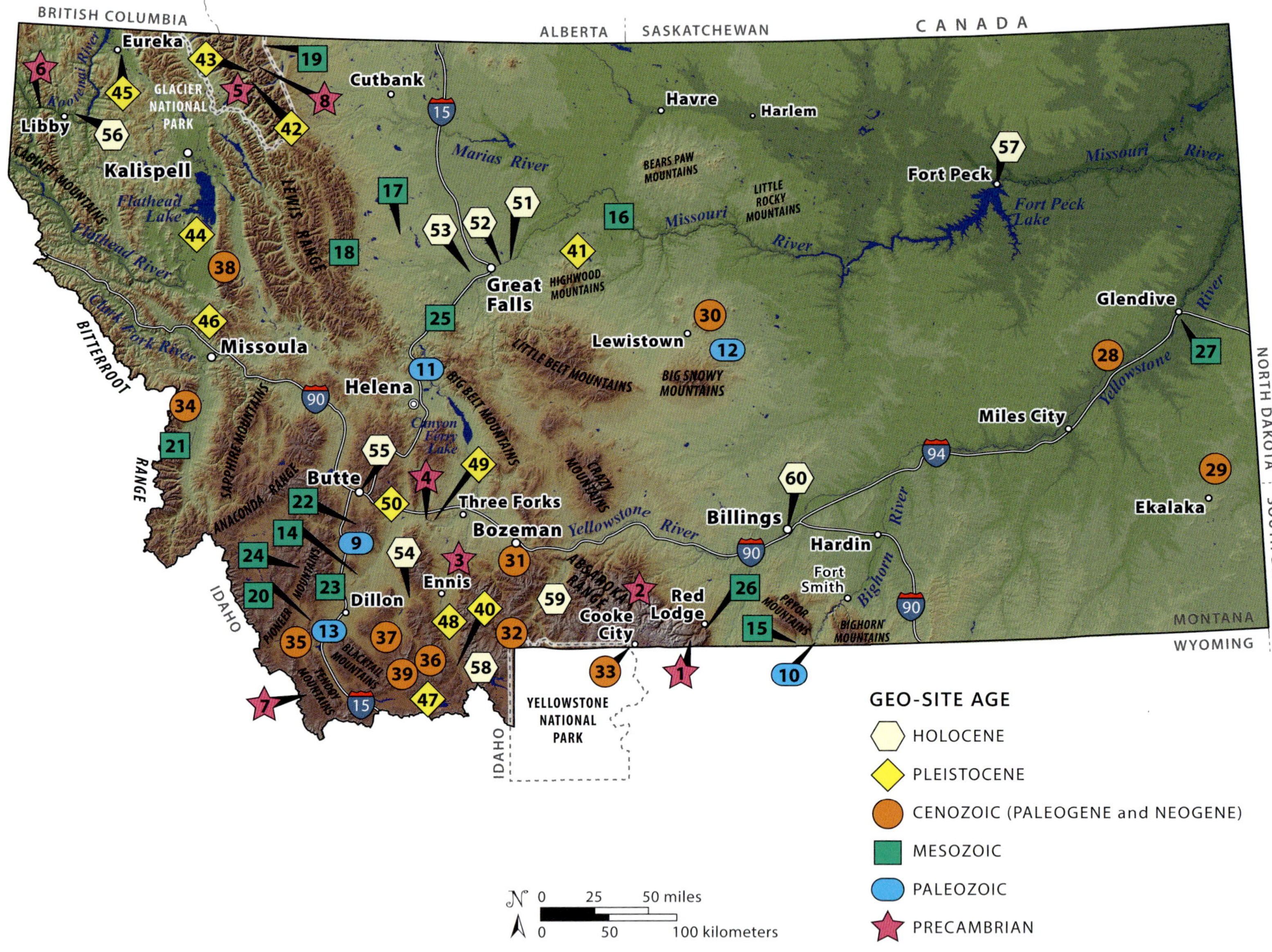
BRITISH COLUMBIA
ALBERTA
SASKATCHEWAN
CANADA
Eureka
Libby
GLACIER NATIONAL PARK
Kalispell
Cutbank
Havre
Harlem
Fort Peck
Fort Peck Lake
Missouri River
Marias River
BEARS PAW MOUNTAINS
LITTLE ROCKY MOUNTAINS
CABINET MOUNTAINS
Flathead Lake
Flathead River
LEWIS RANGE
Great Falls
HIGHWOOD MOUNTAINS
Clark Fork River
BITTERROOT RANGE
Missoula
Helena
Canyon Ferry Lake
BIG BELT MOUNTAINS
LITTLE BELT MOUNTAINS
Lewistown
BIG SNOWY MOUNTAINS
Glendive
Miles City
Yellowstone River
SAPPHIRE MOUNTAINS
ANACONDA RANGE
CRAZY MOUNTAINS
Butte
Three Forks
Bozeman
Billings
Hardin
Ekalaka
NORTH DAKOTA
SOUTH DAKOTA
MONTANA
WYOMING
PIONEER MOUNTAINS
Ennis
Dillon
ABSAROKA RANGE
Cooke City
Red Lodge
Fort Smith
Bighorn River
PRYOR MOUNTAINS
BIGHORN MOUNTAINS
BLACKTAIL MOUNTAINS
TENDOY MOUNTAINS
IDAHO
YELLOWSTONE NATIONAL PARK
15
90
94
GEO-SITE AGE
HOLOCENE
PLEISTOCENE
CENOZOIC (PALEOGENE and NEOGENE)
MESOZOIC
PALEOZOIC
PRECAMBRIAN
0 25 50 miles
0 50 100 kilometers

Contents

GEOLOGIC TIME SCALE

ERA or EON	PERIOD	EPOCH	AGE (mya)	IMPORTANT GEOLOGIC EVENTS IN MONTANA	MAJOR ROCK UNITS	MONTANA GEO-SITES
CENOZOIC	QUATERNARY	HOLOCENE		warming climate; landslides, flooding, earthquakes, and humans modify the landscape; Yellowstone hot spot, 17 mya to present; crustal extension, 50 mya to present	surficial deposits	56 57 58 59 60 51 52 53 54 55
		PLEISTOCENE	0.01	ice sheets advance south into Montana, glaciers form in mountains, and Glacial Lake Missoula fills western valleys	glacial deposits; Huckleberry Ridge Tuff, 2.1 mya	43 47 48 49 50 40 41 42 44 45 46
	TERTIARY: NEOGENE	PLIOCENE	**2.6**	arid climate and pediment formation		
		MIOCENE	5.3	Missouri River flows north off hot spot	Sixmile Creek Formation, about 17–4.5 mya	37 38 39
	TERTIARY: PALEOGENE	OLIGOCENE	23	lake and stream deposits in valleys; plant, insect, fish, and mammal fossils, like camels, horses, and rhinos	Renova Formation, about 48–20 mya	35 36
		EOCENE	33.9	volcanic activity in western and central Montana peak at 50 mya; warm, moist climate and fossil forests	Absaroka Supergroup and Central Montana Alkali Province Volcanics	30 31 32 33 34
		PALEOCENE	56	coastal plain stream deposits and coal; Sevier and Laramide deformation, about 120–50 mya	Fort Union Formation: Tongue River Member; Lebo Member (west), Ekalaka Member (east); Tullock Member	28 29
MESOZOIC	CRETACEOUS		**66**	extinction of dinosaurs; foreland basin; stream and lake deposits (west), marine sandstone and shale in Western Interior Seaway (east), about 100–66 mya; Elkhorn Mountains and Adel Mountains Volcanics; Boulder batholith, about 80–70 mya; Pioneer batholith, about 80–64 mya; Idaho batholith, about 83–53 mya	Beaverhead Group (west), Hell Creek Formation (east); Bearpaw Formation; Two Medicine Formation (west), Judith River Formation (east); Virgelle Formation (west), Eagle Formation (east); Telegraph Creek Formation; Marias River Formation; Blackleaf Formation (west), Thermopolis Formation (east); Kootenai Formation	19 20 26 27; 16 17 18 22 23 24 25; 21
	JURASSIC		145	dinosaur fossils in foreland basin stream deposits; Sundance Sea; oyster and squid fossils; plate collision to the west; Atlantic Ocean begins to open	Morrison Formation; Ellis Group: Swift Formation, Rierdon Formation, Piper Formation	14
	TRIASSIC		201	arid climate, stream deposits, and gypsum; marine limestone with brachiopods; Pangea forms	Chugwater Formation; Dinwoody Formation	15
PALEOZOIC	PERMIAN		**252**	major extinction event at end of Permian; deposition of shelf sandstone, chert, and limestone	Phosphoria Formation	13
	PENNSYLVANIAN		299	arid climate, coastal beaches, and sand dunes; sea-level drop, karst, and red, tropical soil formation	Quadrant Formation; Tensleep Formation; Amsden Group	
	MISSISSIPPIAN		323	Bear Gulch Limestone; rare, soft-bodied marine fossils; tropical marine limestone with corals, crinoids, and brachiopods	Big Snowy Group; Madison Group: Mission Canyon Limestone, Lodgepole Limestone	11 12
	DEVONIAN		359	limestone, dolostone, and shale deposited during rising and falling ocean water across the continent	Three Forks Formation; Jefferson Formation; Maywood Formation	
	SILURIAN		419	no rocks exposed in Montana		
	ORDOVICIAN		444	tropical marine sandy limestone with corals and brachiopods	Bighorn Dolomite	10
	CAMBRIAN		485	Great Unconformity, about 520 mya; first abundant animal fossils in sedimentary rocks deposited when shallow, tropical ocean water transgressed from west to east over eroded basement rocks	Red Lion Formation (west), Snowy Range Formation (east); Pilgrim Limestone; Park Shale; Meagher Limestone; Wolsey Shale; Flathead Sandstone	9
PRECAMBRAIN	PROTEROZOIC: NEOPROTEROZOIC		**541**	breakup of Rodinia Supercontinent, 750 mya	Belt Supergroup: Missoula Group — Pilcher, Garnet Range, McNamara, Libby, Bonner, Mount Shields, Shepard, and Snowslip Formations	7 8
	PROTEROZOIC: MESOPROTEROZOIC		1,000	deposition of Belt Basin sediments, 1,400–1,470 mya; breakup of Nuna Supercontinent (Columbia), 1,600 mya	Belt Supergroup: Piegan Group — Wallace and Helena Formations; Ravalli Group — Empire, Spokane, Grinnell, St. Regis, Revett, and Burke Formations; Lower Belt Group — Prichard, Appekunny, Altyn, and LaHood Formations	4 5 6
	PROTEROZOIC: PALEOPROTEROZOIC		1,600	Big Sky mountain building: collision between Medicine Hat block and Wyoming province, 1,800 mya		3
	ARCHEAN		**2,500**	Stillwater layered intrusion, 2,700 mya; subduction and intrusion of granite, 2,800 mya; oldest basement rock in Montana, 3,600 mya	igneous and metamorphic rocks	1 2

*mya = millions of years ago

A Brief Geologic History of Montana

The mystique of Montana as the Last Best Place is rooted in its natural beauty and unspoiled wilderness. From towering mountains to sweeping prairies, Big Sky Country has captured the imagination of people who visit by the millions each year. The inspirational landscape is a work in progress, billions of years in the making and still changing. Montana's deep-time story is recorded in the rocks, with the oldest rocks and most ancient history at the bottom of the pile, unless disrupted by tectonic forces. Today's landscape is a recent construct, exposing the story for all to read.

Montana's geologic journey begins with the formation of our planet, which came into existence about 4.56 billion years ago from the dust and gas left over from the formation of the Sun. The heat generated by colliding particles and the gravitational pressure on our growing planet created a molten ball that differentiated into a core, mantle, and crust. The outermost layer solidified into the lithosphere, a rigid rind of crust and upper mantle resting on mushy rocks of the underlying asthenosphere. The air was an acrid mixture of water vapor, carbon dioxide, and nitrogen, with no breathable oxygen, and although microbial life may have existed as early as 4.1 billion years ago, our early planet was more like Mars. Most of the early crust was a basaltic heavyweight rich in iron and magnesium minerals. This dense rock sat low on the asthenosphere and collected water condensing from gases, forming the first oceans.

Earth's brittle crust broke into rigid plates driven by movement in the underlying weak asthenosphere. The plates moved and continue to do so at an average rate of a couple of centimeters per year, about the rate at which your fingernails grow. Much of the movement occurs when accumulated stress is rapidly released, and the plates lurch tens of feet causing earthquakes. As plates began to dip beneath other plates, partial melting at depth created magmas rich in silicon and aluminum that then ascended and cooled into granitic rock. The light granitic rocks sat high, making the first significant continents, which grew with time and plate tectonics.

Tectonic action occurs at the boundaries of plates. At divergent plate boundaries, new ocean crust forms as plates spread apart during crustal stretching. When a continent breaks apart, the brittle lithosphere stretches and breaks along parallel faults that drop down basins and uplift ranges. At convergent plate boundaries, plates collide. If oceanic plates are involved, the heavier oceanic lithosphere sinks under lighter continental lithosphere or younger oceanic lithosphere, a process called subduction. The descending slab heats up, dehydrates, and lowers the melting temperature of the overlying rocks, generating magma that rises under pressure and can erupt at the surface as volcanoes. When continents collide, each plate has a similar density, so instead of subducting, the lithosphere crumples, raising

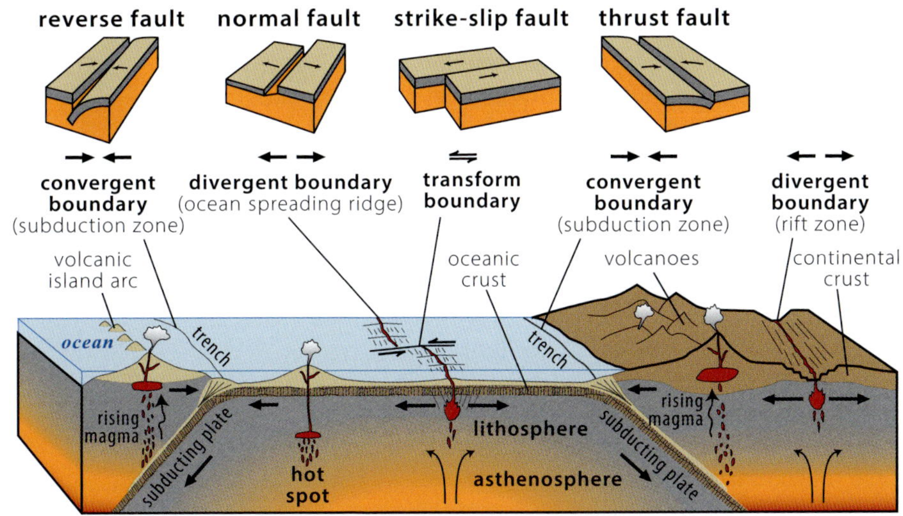

The types of boundaries between tectonic plates at the surface of Earth and the types of faults that are common at each boundary. —Modified from US Geological Survey

high mountains with deep roots. Crustal shortening occurs as rocks are folded and broken on faults that shove older rocks up and over younger rocks. Plates also slide laterally past each other along near-vertical strike-slip faults. Montana has it all, with every plate tectonic process shaping our story in deep time.

As continents formed, the early atmosphere and oceans weathered and eroded them to make sediment. The particles accumulated as horizontal layers of sedimentary rocks, with the oldest layers at the bottom of the pile. If we now find these layers tilted or folded, the deformation occurred after the sediment accumulated. Layered rocks tend to be laterally continuous, meaning you can follow layers in all directions until they thin and change into another type of layered rock or are cut by igneous rocks or faults. Any rock particle contained within another rock is older, and if rocks or faults cut across other rocks, they are younger than the rocks they cut.

The sequence of rocks in any given place on the planet represents only a part of Earth's long history because in no one place is there a continuous accumulation of sedimentary rock. The geologic record is riddled with gaps called unconformities, and the gaps can represent more time than the geologic record. The unconformities usually formed because the land was being eroded. They occasionally occurred because no sedimentary rock was accumulating. Regardless, what we know about the geological story of Montana comes from the gaps in the record as much as from the record itself. Our Montana story is reconstructed from piecing together all the rock sequences studied from every corner of our planet because our human-drawn boundaries are meaningless to the natural processes that govern the planet.

The relationships between rocks can tell geologists which rock is older, but they don't provide numerical ages. Isotopes or unstable elements in minerals in the rocks, however, contain isotopic clocks that start ticking when unstable atoms reach certain temperatures as liquid rocks crystallize into igneous rocks or heated rocks recrystallize into metamorphic rocks. Earth evolves or changes over a span of time so long that geoscientists refer to it as deep time. Put into the context of a single year, with Earth's formation 4.6 billion years ago on January 1, the formation of the first continental crust in Montana begins around mid-March. Animals first appeared in oceans in May, and plants emerged onto land in late November. The Rocky Mountains were first raised in mid-December, about the same time that granitic magma injected into older rocks, depositing copper, silver, and gold that would one day attract miners. Dinosaurs began roaming the coastal plain of a great interior seaway at this time, only to meet their demise before the end of the month. Explosive volcanoes erupted in late December, and the Basin and Range topography we see today began forming on the evening of December 31. Glacial ice last covered the mountains and plains around one minute and thirty seconds before midnight, just as humans began hunting wooly mammoths and driving bison over cliffs. The miners were the last on the scene, arriving less than one second before the end of our deep-time year. It's about perspective.

Layered rocks tend to accumulate like dust on a shelf, with the oldest rocks at the bottom of the pile and the youngest at the top, like these lake deposits in the Cretaceous Adel Mountains volcanic field near Craig, Montana.

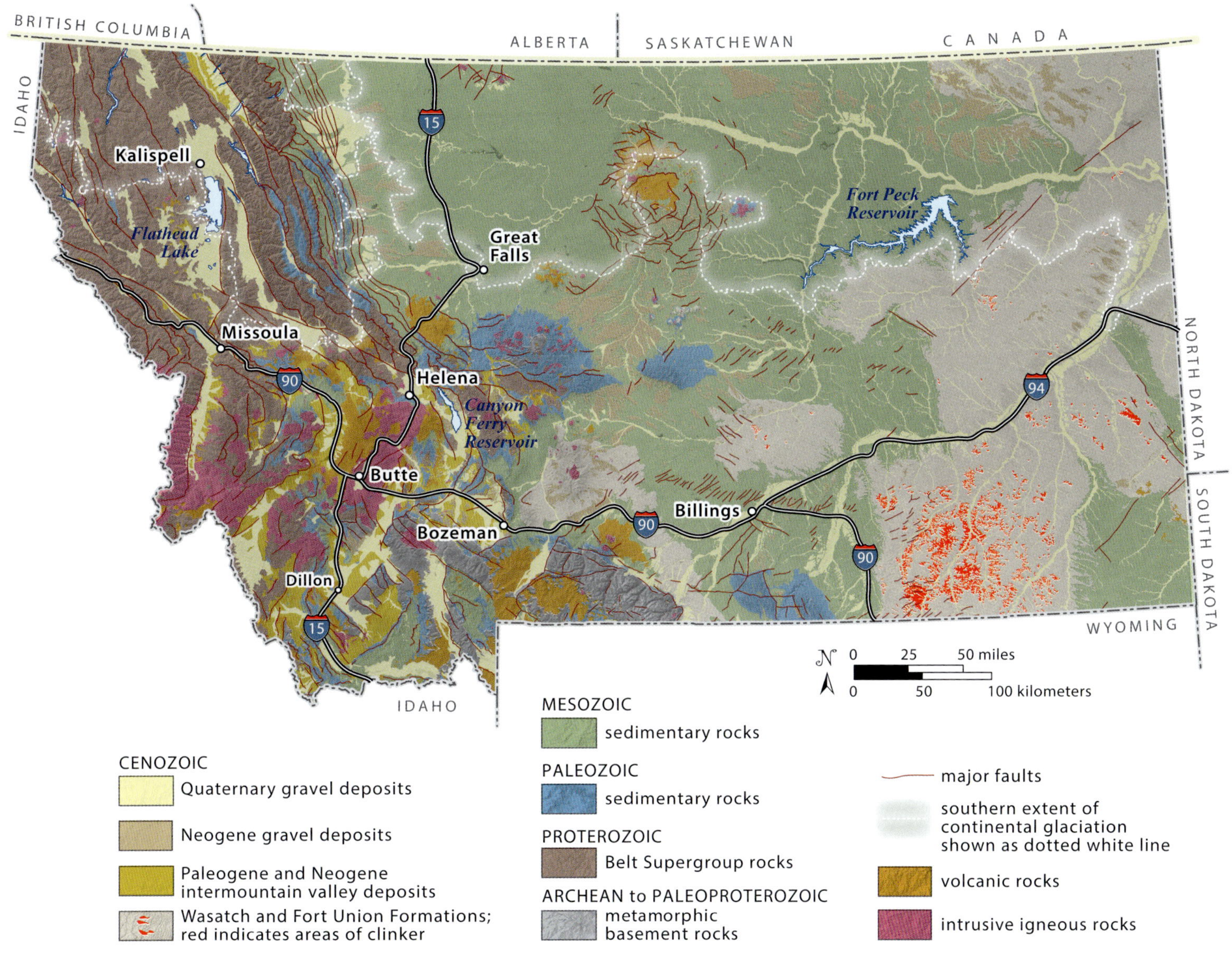

The simplified geologic map of Montana with rocks grouped by ages and types. —Modified from the Montana Bureau of Mines and Geology

Precambrian Era

Rocks are like people—they tend to become more metamorphosed with age. In Montana, rocks from the basement of time are very old, ranging from 3.6 to 2.55 billion years old. Some of these rocks contain crystals eroded from even older rocks that formed 4.0 billion years ago, making them among the oldest minerals on Earth. Montana's old rocks began as ancient sedimentary and igneous rocks that were heated and stretched at high pressures and temperatures to form schist and gneiss, streaky metamorphic rocks formed during the birth of Montana's first continent, the Wyoming province. A massive outpouring of granitic magma, now granitic gneiss, around 2.8 billion years ago shows that plate subduction had started. We call these old rocks the basement because they underlie most rocks across the state but are only exposed in a few mountain ranges. They remain mostly buried below a thick pile of sedimentary and volcanic cover in the western valleys and on the plains.

The next 2 billion years in Montana began with the deposition of oceanic sediments along the western margin of the Wyoming province, including banded iron formations that record the slow oxygenation of the planet by photosynthetic bacteria. Ocean floor subduction scraped off and added the sedimentary rocks to the province and brought along a small continent, called the Medicine Hat block, that was joined to the Wyoming province about 1.8 billion years ago during the

The oldest rocks in Montana are metamorphic and igneous rocks of Archean age, exposed here in Yankee Jim Canyon near Gardiner.

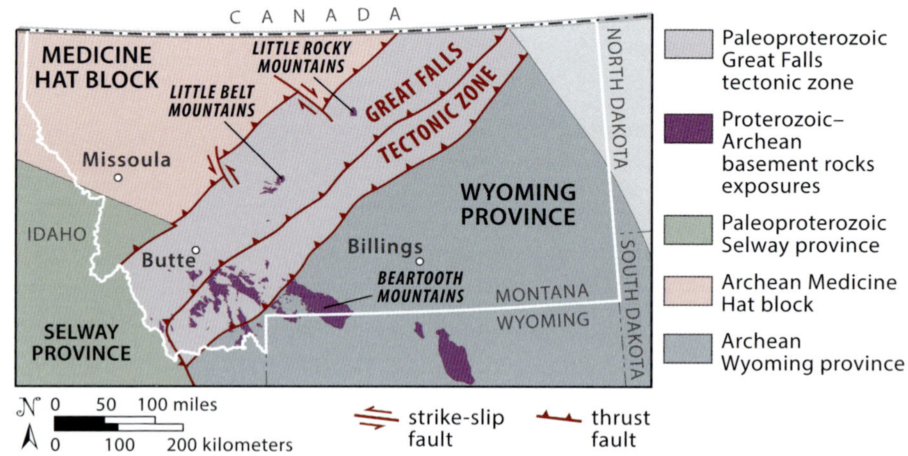

The area where the Medicine Hat block and the Wyoming province joined in the Paleoproterozoic is called the Great Falls tectonic zone. It contains rocks intruded and metamorphosed during the Big Sky mountain building about 1.8 billion years ago. The dark purple areas are the places where Archean and Paleoproterozoic rocks are exposed at the surface today.
—Modified from Foster and others, 2006

Big Sky mountain building episode. The continental collision completed the assembly of the continental lithosphere in Montana and raised Himalayan-sized mountains. When it was over, the Wyoming province was officially part of the continental core of North America, a continent called Laurentia.

The North American continental core was one of many that joined together to form a supercontinent called Nuna (also called Columbia), and the amalgamation of continents began to break apart around 1.5 billion years ago. In Montana, stretching apart of the supercontinent opened the Belt Basin, a lowland where sediment accumulated. For about 70 million years, the Belt Sea waxed and waned, accumulating nearly 60,000 feet of mostly fine-grained sand, mud, and limestone. Sediment that became trapped in mats of photosynthetic cyanobacteria created distinctive, domal-shaped structures called stromatolites. No multicellular life existed on Earth to churn up the sediments, so the Belt Supergroup rocks are famous among geologists for their exquisite sedimentary structures, like mud cracks, ripple marks, and raindrop imprints. Around 800 million years ago, yet another supercontinent, Rodinia, broke apart, splitting the Belt Basin and forming a new ocean coastline to the west of Montana.

Stretching of the crust around 1.5 billion years ago opened the Belt Basin, which subsided and filled to a depth of more than 10 miles with fine-grained sediment, like these layers of fine-grained sandstone covered with ripples on Flint Creek Hill south of Philipsburg.

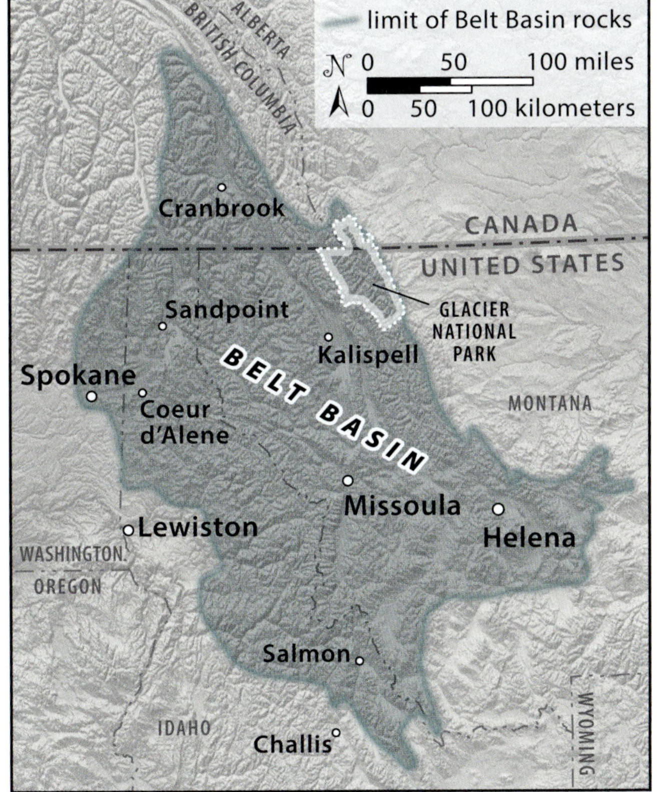

Approximate outline of the Belt Basin based on the distribution of Belt Supergroup rocks. —Modified from Winston, 1989

Paleozoic Era

More than a billion years of erosion left most of Montana a featureless plain by the start of Paleozoic time, about 541 million years ago. A rising ocean from the west reached Montana by 520 million years ago in the first period of the Paleozoic Era (the Cambrian), spreading beach sand over the eroded surface of metamorphic rock and leaving a great unconformity in the rock record. Continued sea-level rise brought shelf mud and eventually offshore limestone, the first of many tropical marine deposits from the Cambrian seas inundating the continent. The Cambrian rocks are full of the shells of trilobites, brachiopods, and other tropical marine animals that exploded in diversity at the beginning of Paleozoic time. For the next 270 million years, the sea rose and fell, and the continent subsided under the weight of the sediment. As more sediment accumulated, it stacked up into a thick pile of marine and nonmarine sedimentary rocks. These years saw the first vertebrate animals in the oceans and on land, as well as the first land plants. At the end of the Paleozoic Era, the last great supercontinent formed. Known as Pangea, it contributed to a changing climate that likely caused the greatest mass extinction in the history of life on Earth.

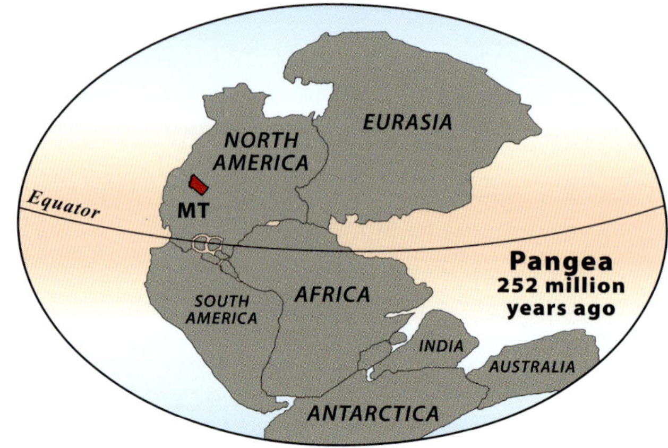

The supercontinent Pangea.

Shallow, tropical oceans covered Montana during much of Paleozoic time, depositing the layers of limestone exposed in the Bighorn Canyon south of Billings.

A fossil bryozoan from Paleozoic limestone in southwest Montana. These filter-feeding animals lived in clear, tropical marine water that covered much of the continent more than 300 million years ago. The specimen is about 2 inches long.

Mesozoic Era

Supercontinents tend to be short lived, so when Pangea began breaking up and the Atlantic Ocean opened in Mesozoic time, the westward-moving North American plate pushed into the oceanic lithosphere and small volcanic islands to its west. The western edge of the continent became a convergent plate boundary beginning about 160 million years ago. In what is called the Sevier mountain building episode, the earliest rendition of the Rocky Mountains was raised as the continental edge was compressed like an accordion. A thick sedimentary pile was crumpled and shoved tens of miles eastward, forming a wedge-shaped feature called a fold-and-thrust belt, with older rocks shoved over younger rocks. This type of crustal shortening, with faults at low angles to Earth's surface and moving mostly sedimentary rock and not much deep basement rock, is called thin-skinned deformation. The thickened crust generated granitic magma that fed volcanoes and intruded into the actively deforming fold-and-thrust belt.

The weight of all that thick crust in western Montana depressed the crust to the east, forming a north-south trough or foreland basin that was inundated by seawater of the Sundance Seaway in Jurassic time, the middle of the Mesozoic Era. Marine sediments accumulated, some loaded with the fossils of oysters and squid-like animals called belemnites. The ocean retreated by the end of Jurassic time, around 145 million years ago, and stream sediments began eroding off the rising mountains of the fold-and-thrust belt to the west, spreading across the foreland basin. The accumulated sediment buried peat, converting it to coal, and also the remains of sauropod dinosaurs, the largest land animals to ever live. By Late Cretaceous time, around 100 million years ago, ocean water returned to the foreland basin to create the Western Interior Seaway, an inland sea extending from the Arctic Ocean to the Gulf of Mexico. Dinosaurs roamed the coastal plain, and the seaway was ruled by swimming reptiles and squids called ammonites.

The seaway waxed and waned in Cretaceous time, the last period of the Mesozoic Era, resulting in a thick stack of interbedded sandstone and shale as the foreland basin subsided under the weight of the sediment. The fold-and-thrust belt

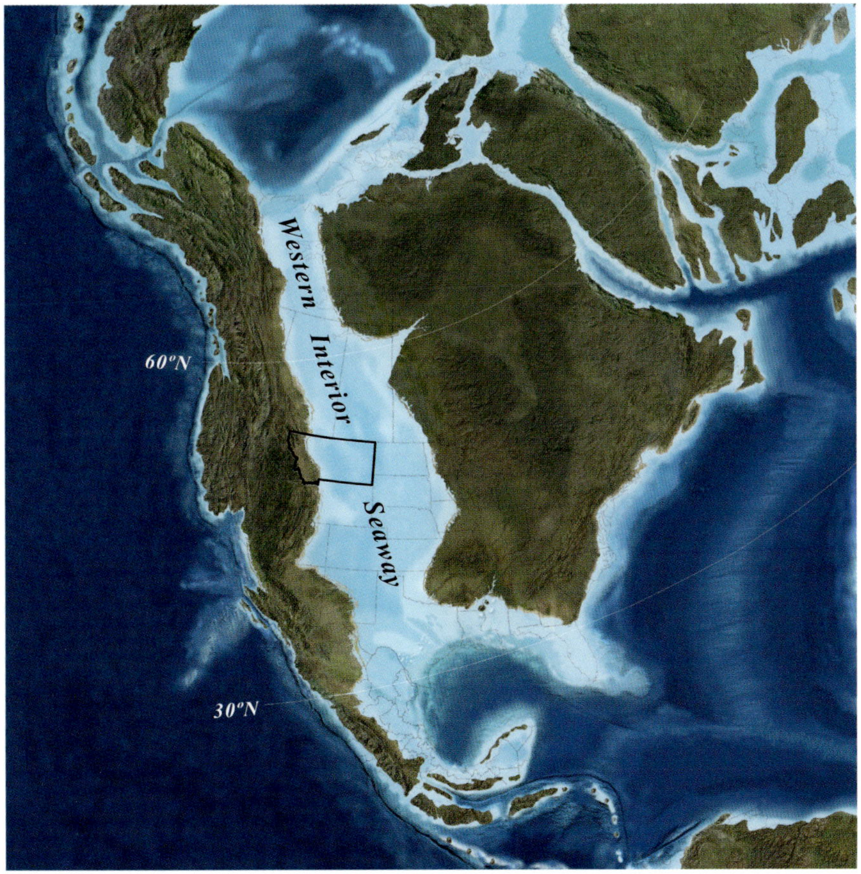

In Late Cretaceous time, around 75 million years ago, the Western Interior Seaway connected the Arctic Ocean with the Gulf of Mexico and separated North America into two landmasses. —Image © 2023 Colorado Plateau Systems

remained active, slowly migrating from west to east, and deforming some of the foreland basin sediments shed off the rising mountains. By the end of Mesozoic time, the thin-skin Sevier mountain building was joined by a thick-skin Laramide mountain building during which faults at a higher angle to Earth's surface broke the underlying basement rocks and shoved up huge blocks of it to form mountains. It is called thick-skinned because it involved a much thicker portion of Earth's crust. The overlying Paleozoic and Mesozoic sedimentary rocks were draped over the rising basement and later mostly eroded, leaving steeply dipping palisades of sedimentary rocks along the fronts of mountain ranges composed of crystalline basement rocks.

Although the Laramide style of deformation is overall a bit younger and dominant to the east, the two styles of crustal shortening overlap in time and space. The reason for the change in structural style is a matter of great debate and is not yet determined. For many years, geologists argued

Intensely folded and faulted layers of Cretaceous Kootenai limestone at Block Mountain north of Dillon.

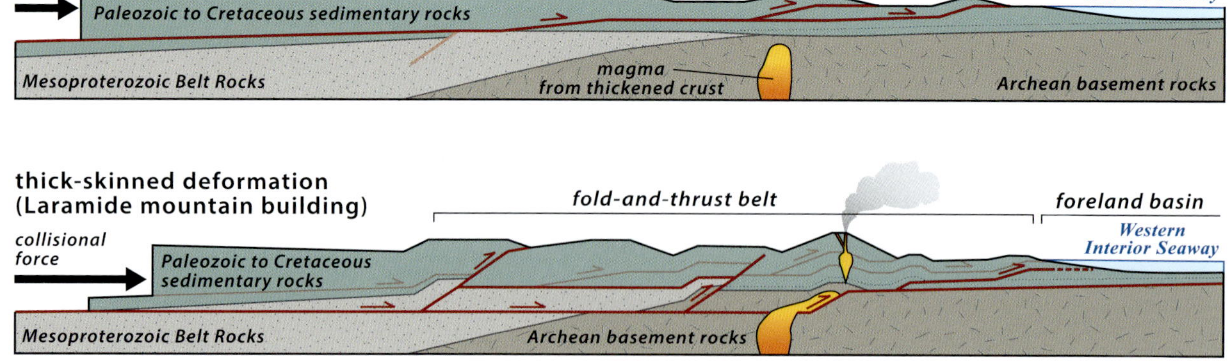

Over time, the deformation in Montana shifted from thin-skinned to thick-skinned basement blocks, like the Beartooth uplift near Red Lodge. —Modified from Parker and Pearson, 2021

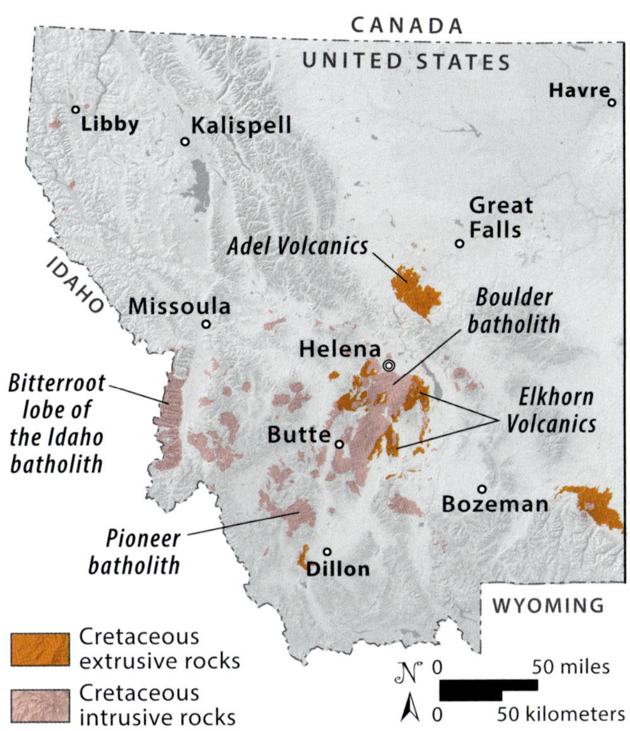

Distribution of Cretaceous intrusive and extrusive igneous rocks in Montana. —Modified from Hearn and others, 1989

Cretaceous and Paleogene granitic rocks in Blodgett Canyon west of Hamilton.

that a transition from steep to shallow subduction caused drag against the North American plate and popped up the basement-cored uplifts. In more recent models, a complex tectonic collision with large landmasses first produced the Sevier thrusting, then transitioned to Laramide deformation as the landmasses moved north. No matter the cause, there can be no doubt about the importance of this time to the state's economics. The deformation, magmatic fluids, and sedimentation associated with these processes formed most of the ore minerals, coal, and oil and gas deposits and buried the remains of the world's most famous and economically valuable dinosaurs.

Tyrannosaurus upper jaw at the Montana Dinosaur Center. Quarter for scale.

Thousands of dark-colored, coarse-grained dikes like this one radiate out from the Cretaceous Adel Mountains volcanic field near Craig.

A spectacular fold exposed upstream from Yellowtail Dam in the Bighorn Reservoir south of Hardin. These folded layers drape over a basement block that was shoved up during the Laramide mountain building.

Cenozoic Era

The beginning of the end of our story was a tumultuous time of continued tectonic upheaval and biological crises. The Western Interior Seaway retreated from Montana, and streams chased it eastward, depositing sediments eroded from the western mountains onto the coastal plain. The sand and clay buried peat bogs that now form large economic coal deposits in southeastern Montana. The dinosaurs perished when a giant rock from space struck Earth 66 million years ago, opening the door for mammals to flourish. The western edge of the North American plate was so far to the west that it is hard to understand how Montana continued its volcanic and mountain building dance, yet igneous activity became even more widespread, forming dome-shaped intrusions—called laccoliths—of strange compositions in central Montana, granitic intrusions in the southwest, and explosive stratovolcanoes over a large part of central Idaho, southwest Montana, and western Wyoming.

The chaos peaked around 50 million years ago, about the time crustal stretching began bringing deep rocks to the surface along low-angle faults, and big valleys filled with sediments eroded off the rising domes. It is unclear what caused so many different tectonic and volcanic events to happen at the same time. Perhaps deep rocks decompressed and melted when the fold-and-thrust belt collapsed as the convergence to the west lessened, like a bulldozer backing off a pile of dirt. We know the Eocene climate was warm and wet because volcanic mudflows buried forests of dawn redwood trees.

Volcanism continued into Oligocene and Miocene time, but the composition was now becoming more basaltic, changing from the more silica-rich granites and rhyolites of Eocene time. The silica-rich magmas melt from the crust whereas silica-poor basaltic magma melts from the mantle. Some geologists interpret the chemical change as evidence that a subducting plate was progressively sagging into the mantle, causing extension in the overlying crust that ultimately allowed basaltic magma to rise directly from the mantle. Tectonic stretching also continued, forming basins along a trend of the former Sevier thin-skin shortening, but creating a large, stable basin in and around eroded remnants of the Laramide uplifts. The basins filled with alluvial fan, stream, and lake sediments of the Renova Formation, preserving world-class fossils of mammals, fish,

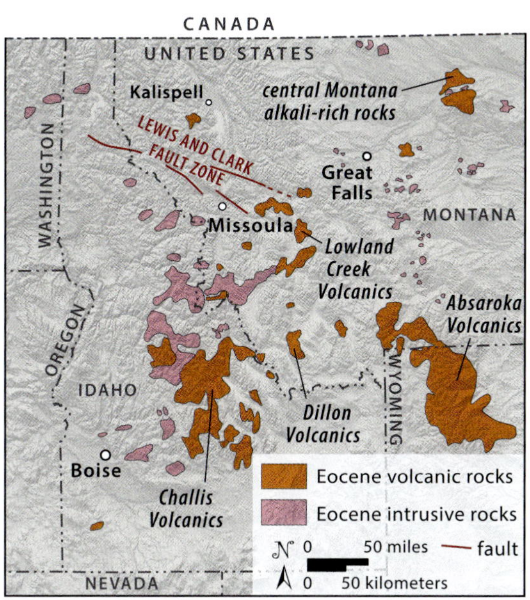

Distribution of Eocene-age intrusive and extrusive rocks. —Modified from Hyndman and Thomas, 2020

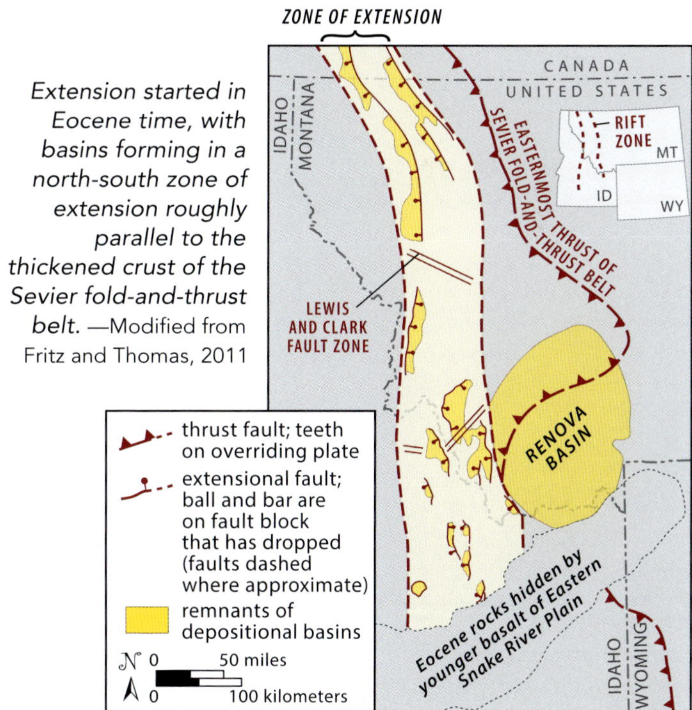

Extension started in Eocene time, with basins forming in a north-south zone of extension roughly parallel to the thickened crust of the Sevier fold-and-thrust belt. —Modified from Fritz and Thomas, 2011

insects, and plants that record a drying climate into Miocene time in the late Cenozoic.

Fossil insect, about 0.5 inches long, from lake deposits of the Oligocene-aged Renova Formation near Ruby Reservoir in southwest Montana.

Around 17 million years ago, regional changes in plate motions caused crustal stretching in the Intermountain West of the United States, breaking the lithosphere into north-trending basins and ranges. In western Montana, the older basins continued to deepen, while new basins formed that accumulated alluvial fan and stream deposits of the Sixmile Creek Formation. The Yellowstone hot spot began forming calderas in northern Nevada at this time, heating the crust and causing it to expand or bulge upward. Northeast-trending basins formed in southwest Montana as the crust stretched to accommodate the thermal bulging. The ancestral upper Missouri River followed these basins, transporting ash and gravel off the thermal bulge into Hudson Bay from 17 to 4.5 million years ago. As the westward movement of the North American plate saw the hot spot situated closer to its present position under Yellowstone National Park, the thermally raised crust collapsed and broke into northwest-trending extensional basins and ranges that changed the path of the ancestral Missouri River. Deposits of the ancient river now reside high in the northwest-trending mountains of southwest Montana, and the faults are active and dangerous, as was shown in 1959 by the 7.3 magnitude temblor near Hebgen Lake (site 58).

Some mountain ranges were eroded back by flash flooding during Pliocene time, from 5 to 2.6 million years ago.

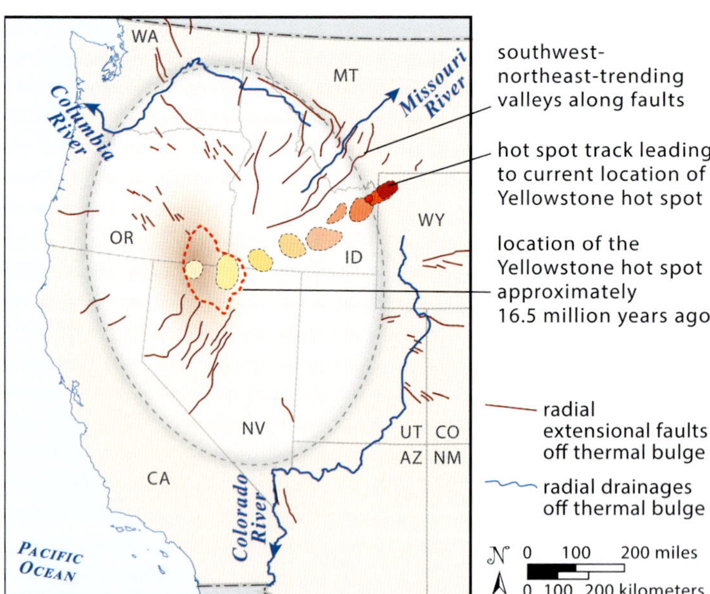

Path of the ancestral Missouri River off the uplifted bulge created by heat of the Yellowstone hot spot starting around 17 million years ago. —Modified from Sears and others, 2009

This nearly 100-foot-thick ash deposit of the Sixmile Creek Formation in the Sweetwater Range near Dillon was transported here by the ancestral Missouri River, which carried incredible volumes of ash during volcanic eruptions from the Yellowstone hot spot.

The erosion formed flat, sloping pediment benches that rise above the modern drainages. The wet climates that came with the Pleistocene ice ages at the end of the Cenozoic filled rivers, which then cut the modern stream valleys. In Montana, there is evidence for at least two ice advances, one that peaked around 140,000 and another around 20,000 years ago. The alpine ice sculpted mountains into the majestic severity that attracts people by the millions to Glacier National Park. On the plains, the huge Laurentide ice sheet flowed down from Canada, diverting the Missouri River to its present path and leaving behind glacial debris, lake deposits, and outburst flooding channels. A lobe of the Cordilleran ice sheet, combined with alpine glaciers from the mountains, flowed down the Flathead and Mission Valleys, leaving glacial debris that impounded Flathead Lake (site 44). A lobe of ice blocked the Clark Fork River in northern Idaho, creating Glacial Lake Missoula (site 46), which filled and spilled at least forty times, flooding eastern Washington with biblical torrents.

Alpine glaciers sculpted the peaks in Glacier National Park.

The First People navigated the severe climate of the last ice age, hunting mammoths and using the natural landscape to run bison off cliffs. Most of the ice retreated from Montana by Holocene time, about 11,700 years ago. Migrants seeking their fortune in gold arrived in the 1860s, constructing settlements that are now our cities. Modern society faces many challenges that could be better managed by understanding the deep-time story presented in this book. From earthquakes that might destroy our vulnerable historic cities to the contamination caused by past mining to our proximity to the world's largest explosive volcanic center, geology is always at bat. Perhaps our biggest challenge is a changing climate compromising water resources and putting into question the very future of the Last Best Place.

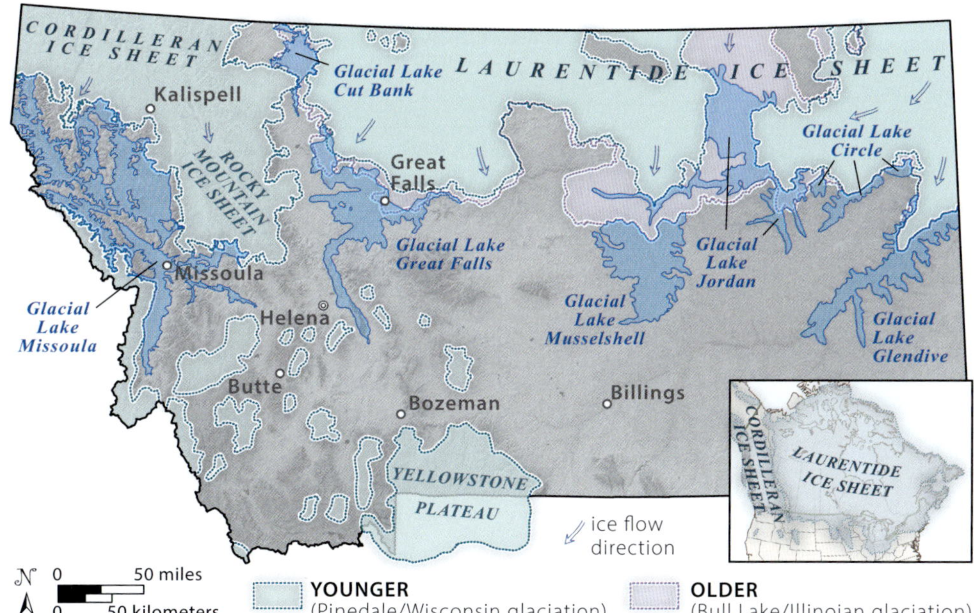

The approximate maximum extent of the Illinoian (Bull Lake) glacial ice and Late Wisconsin (Pinedale) glacial ice and glacial lakes formed during the last glacial maximum around 20,000 years ago. —Modified from Fullerton and others, 2012

Beartooth Highway

Montana's Oldest Rocks

Rocks exposed at the Rock Creek Vista, near the top of the switchbacks on the Beartooth Highway south of Red Lodge, are some of the oldest in Montana. The metamorphic rocks along the quarter-mile trail to the overlook range in age from 3.6 to 2.8 billion years and preserve the crust-forming processes that built Montana's first continent, the Wyoming province. It is one of several small landmasses that were welded together during the initial construction of the larger continent of North America, called Laurentia. Most rocks along the trail are gneiss, composed of the minerals quartz, feldspar, and mica distributed into light and dark bands called foliation. The gneiss began as igneous or sedimentary rock but was metamorphosed and folded by the intense heat and pressure caused by the assembly of the Wyoming province. Some rocks preserve dark blobs of older metamorphic rocks, while others show that the rocks melted, forming mixtures of igneous and metamorphic rocks called migmatite. Dikes, or sheets of igneous rock that form in the fractures of existing rock bodies, cut across the gneiss bands. Some of the dikes are as old as the rocks they injected, with large crystals called pegmatite. Other dikes are composed of small, white feldspars that crystallized a mere 50 million years ago. See if you can find them along the trail.

This place is designated a global Geological Heritage Site by the International Union of Geological Sciences because the rocks are valuable for understanding the evolution of Earth's early crust. Some of the rocks contain zircon crystals eroded from even older rocks that formed 4.0 billion years ago, making them among the oldest minerals on Earth. The ancient zircon crystals may have originally crystallized in the earliest basaltic crust. The birth of granitic continents began about 3.6 billion years ago, marking the onset of plate tectonic subduction. Voluminous granitic magma, now granitic gneiss like some of the rocks exposed along the trail, intruded the older rocks around 2.8 billion years ago.

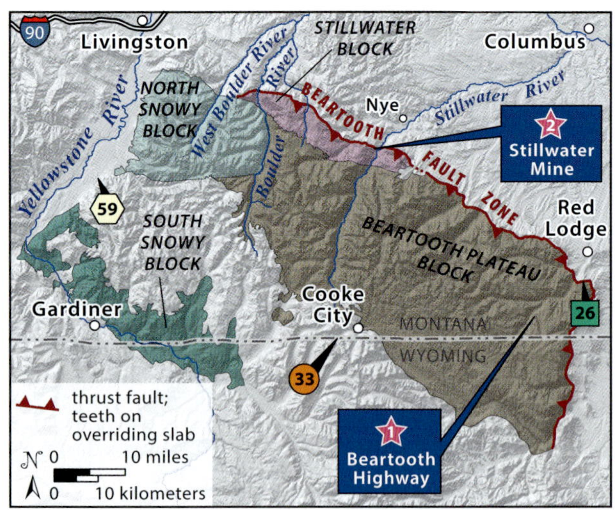

The Beartooth Plateau, several blocks of Archean rock (shown in color), were uplifted along the Beartooth thrust fault. The Stillwater Complex is the subject of Site 2.

From Red Lodge, travel 20.6 miles south on US 212 to Rock Creek Vista. Park and walk about a quarter mile to the end of the vista trail.

View southwest from the Rock Creek Vista of the planar surface of the Beartooth Plateau and U-shaped glacial valley eroded into its hard Archean rocks.

Archean rocks underlie younger rocks across nearly all of Montana, but these old rocks are mostly exposed in southwest Montana, especially in the Beartooth uplift. From 65 to 57 million years ago, this huge basement block with its cover of younger rocks was raised tens of thousands of feet along steep, compressional faults formed when landmasses and ocean floor to the west collided with the North American continent during the Laramide mountain building. Erosion of the younger rocks exposed the old metamorphic rocks, which had already been deeply eroded to a flat surface by the end of Precambrian time. That surface persists to this day in the relatively flat Beartooth Plateau, which is clear to see to the west from the Rock Creek Vista.

Folded granitic gneiss up to 3.6 billion years old exposed on the Rock Creek Vista trail.

2 Stillwater Mine

A Large Layered Intrusion

The Wyoming province continued to evolve and grow westward during the latter part of Archean time with the addition of metamorphic and igneous rocks, like the Stillwater Complex along the Beartooth Mountain front between Livingston and Red Lodge. These iconic, layered igneous rocks were intruded into the crust around 2.7 billion years ago at depths of 6 to 10 miles. The sheet-like intrusion is at least 4 miles thick, even though some of its top was eroded off prior to burial by younger sediments in Cambrian time. During the Laramide mountain building, it was tilted on edge with the rising basement of the Beartooth uplift and is now exposed for 30 miles along the northern mountain front. The rocks are so geologically important, they were designated by the International Union of Geological Sciences as a global Geological Heritage Site.

The most notable place to see the rocks is on National Forest Road 846 above the Stillwater Mine west of Nye. Much of the base rests on baked Archean sedimentary rocks, and the complex is divided into three series, from bottom to top: Basal, Ultramafic, and Banded. They are composed of peridotite and gabbro, intrusive igneous rocks composed of iron- and magnesium-rich, silica-poor minerals like olivine, pyroxene, and calcium plagioclase. The Ultramafic series contains chromite-rich seams mined during World War II for making steel, and the world's highest-grade platinum group deposits occur in the lower Banded series in a thin zone called the J-M Reef. It's the largest source of these elements in the United States, and catalytic converters are the primary market. The Banded series boasts rare, light-colored rocks called anorthosite that are common on the moon and so were used to train astronauts for identification.

Intrusive igneous rocks are typically not layered because most viscous, granitic magmas crystallize within Earth's crust in relatively small magma chambers that cool evenly, allowing minerals to crystallize randomly in all directions. In contrast, large, fluid magma chambers, like that of the Stillwater Complex, are more likely to allow crystals to settle and separate based on density, creating distinct layers of minerals that extend laterally for miles. The details are debated, but layering in the Stillwater Complex likely formed through fractional crystallization, a process where minerals crystallize at different temperatures in slowly cooling magma, separate from it, and form distinct layers in the magma chamber. Heavy minerals like olivine formed first and sank to the bottom, followed by pyroxene and plagioclase feldspar.

The magma that helped form the Stillwater Complex likely came from melting of Earth's mantle, where most iron- and magnesium-rich rocks originate today. Perhaps it was a time of crustal stretching, which reduced pressure on the underlying mantle, causing it to partially melt and form ascending magmas that pooled within the continental crust.

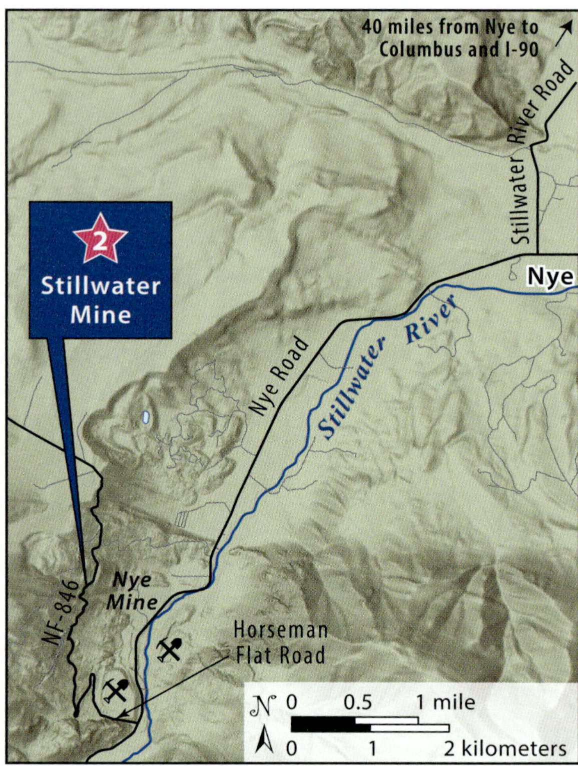

From Nye, travel 5.4 miles south on Nye Road, and turn right (west) on Horseman Flat Road. At 0.6 mile, turn left (west) on NF-846 for 1.5 miles to a roadcut at 45.3964, -109.8822.

Alternating layers of dark-colored pyroxene and white plagioclase in the tilted Banded series rocks of the Stillwater Complex on National Forest Road 846 above the Stillwater Mine.

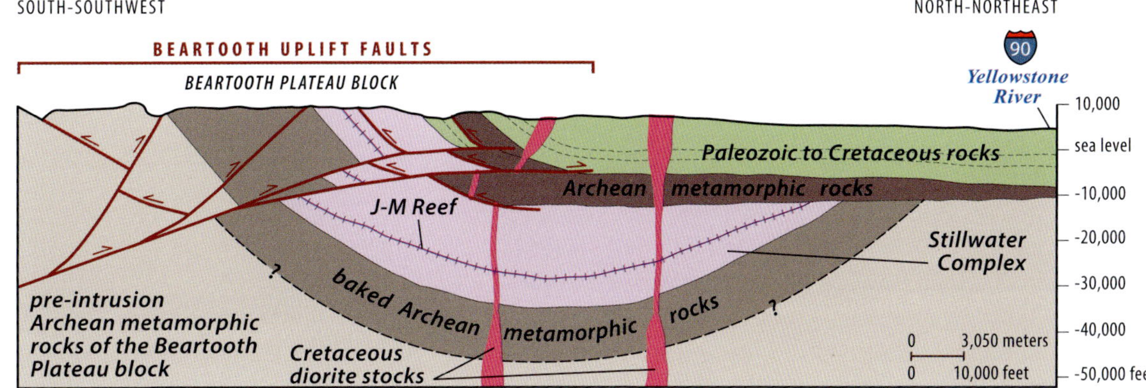

Geologic cross section of the Stillwater Complex. —Modified from Boudreau and others, 2020

Bear Trap Canyon

Metamorphic Rocks Formed by Plate Collisions

During Paleoproterozoic time, plate collisions added ocean floor rocks and small landmasses to the western margin of the Wyoming province. Among the rocks are metamorphosed ocean floor basalt and sedimentary rocks metamorphosed to quartzite, marble, schist, and gneiss. It was all then added to the North American continent (Laurentia) around 1.8 billion years ago. The metamorphic rocks are now exposed in uplifted mountain ranges in southwest Montana, with good access in Bear Trap Canyon, a gash in the rocks carved by the Madison River.

The old sedimentary rocks are mangled almost beyond recognition, but they still provide clues about the maturing surface of the planet. At the time of their deposition, more than 2.5 billion years ago, continents were barren of life, but the oceans teemed with oxygen-producing bacteria, called cyanobacteria, which oxygenated the oceans and atmosphere. During the Great Oxidation Event, fluctuating oceanic oxygen levels produced banded iron formations, rocks with alternating layers of iron oxide minerals and quartz. These banded rocks are found along the trail not far from the parking area. The marble formed by bacterial photosynthesis, which depleted carbon dioxide in the surrounding water, reducing its acidity and precipitating limestone. Some rocks preserve remnants of the original sedimentary processes, like cross-bedding, which has internal layers at an angle to the main layers. The inclined layers formed as grains of sand bounced along the bottom of a coastal stream or beach billions of years ago. Earth was different yet vaguely familiar.

Studies show the rocks in Bear Trap Canyon were deformed several times by the heat and pressure of plate collisions that buried them to depths of more than 25 miles. Such events can heat rocks to 1,500°F and crush them at pressures up to 10 kilobars or 145,000 pounds per square inch (psi). For reference, a typical car tire is inflated to 32 psi. The intense heat and pressure slowly transformed the original rocks, aligning the minerals into parallel layers called metamorphic foliation. When the pressure direction changed, perhaps due to another plate collision, the older foliation was overprinted and folded.

The oldest of the Paleoproterozoic plate collisions on the northwestern margin of the Wyoming province occurred 2.45

From Norris, travel 9 miles east on MT 84, and turn right on South Beartrap Canyon Road. Follow the dirt road for 3.2 miles to Bear Trap Canyon Access at 45.5774, -111.5953.

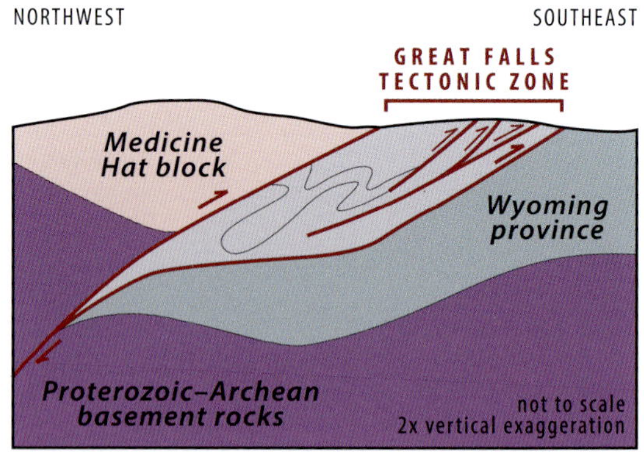

Simplified cross section of the Big Sky mountain building. —Modified from Harms and others, 2006

Complex folding in metamorphic rocks in Bear Trap Canyon. The parallel layering is metamorphic foliation.

billion years ago and is poorly understood. Any mountains created were eroded away, and by 2.0 billion years ago, the area was stretched apart, and dikes intruded. The last plate collision, called the Big Sky mountain building, occurred from about 1.8 to 1.7 billion years ago and records the joining of volcanic islands and a small landmass called the Medicine Hat block to the western edge of the Wyoming province. The northeast-trending suture zone, called the Great Falls tectonic zone, is composed mostly of rocks metamorphosed during the event and igneous rocks formed by subduction as the ocean basin between the continents closed. By the time it was over, the Wyoming province firmly connected to Laurentia, completing the assembly of the old core of North America.

Jefferson River Canyon

The Southern Edge of the Belt Basin

During the addition of the Wyoming province to Laurentia around 1.8 billion years ago, continents around the globe were assembling into a supercontinent, called Nuna by some geologists and Columbia by others. The supercontinent started to fragment 1.5 billion years ago, about the time the crust in Montana began stretching apart and subsiding to form a gigantic continental basin of sediment accumulation called the Belt Basin. Signs of the continental break-up include dark-colored diabase sills, which were magma from the mantle that intruded between layers of the basin's oldest sediment. Some of the oldest deposits in the basin are exposed in the Jefferson River Canyon east of LaHood, where the Jefferson Canyon fault defined the basin's southern margin.

The canyon's western entrance is lined with steep, foreboding cliffs of the LaHood Formation, the oldest sedimentary deposits of the Mesoproterozoic Belt Supergroup. An estimated depth of 10 miles of sediment filled the Belt Basin between 1.47 and 1.40 billion years ago. The Belt Sea occupied the basin, fed by rivers draining Laurentia from at least three sides, including highlands to the south of the Jefferson Canyon fault called the Dillon block. At the time, this fault was active, dropping the Belt Basin down to the north and uplifting mountains to the south, where Archean and Paleoproterozoic metamorphic rocks were exposed, eroded, and transported northward into the Belt Basin. The largest rocks were deposited nearest to the highlands, while the finer sediments were transported farther out into the basin. The accumulating sediment gets finer-grained higher up in the stack of rocks as well, showing a transition over time to environments with slower currents.

In Jefferson River Canyon, the LaHood Formation is mostly a dark-colored sandstone and conglomerate with rocks of various sizes, some as big as cars. They are mostly metamorphic rocks, likely eroded from the Dillon block. Some float in a matrix of sand-sized grains, likely deposited in muddy mixtures of rock and water called debris flows. Others are cross-bedded, like in river deposits where grains bounced along the bottom. Many rocks are angular, showing they didn't get rounded during transport. Either they didn't travel

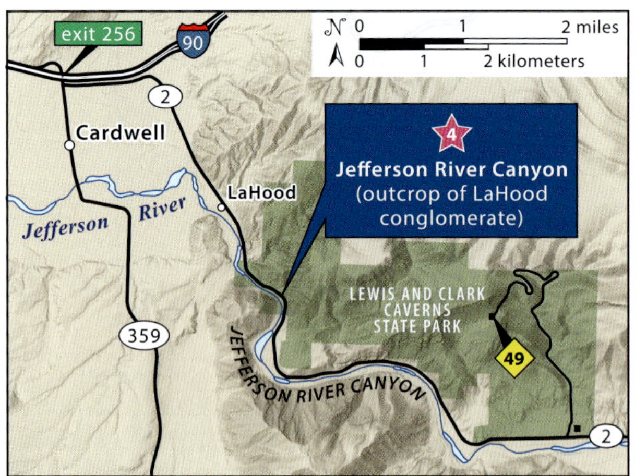

From the Cardwell interchange of I-90 and MT 2, travel 3.5 miles east to a pullout on your left (east) just past a large roadcut.

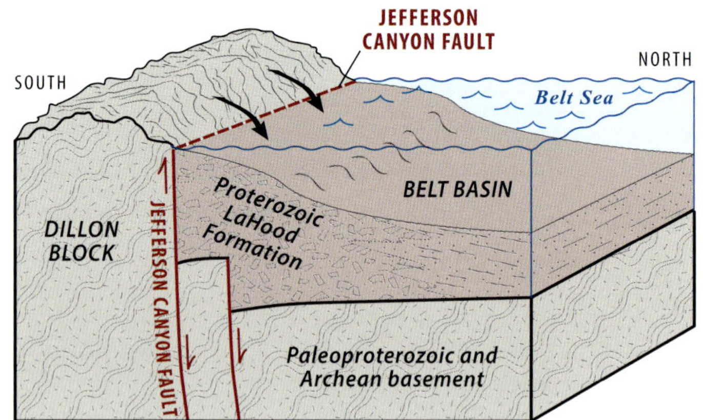

The LaHood Formation was shed northward off the tectonically active Dillon block to the south. —Modified from Hyndman and Thomas, 2020

very far from their source, or they were cushioned during transport by the surrounding sand. Some deposits are not layered at all, like they were rapidly dumped, while others are layered with the largest particles at the bottom, typical of grains deposited as the water current slows.

Such sedimentary processes occur not only on surficial alluvial fans but also on underwater fans, where intact blocks of sediment slowly slump and turbid flows of water and sediment move rapidly downslope by gravity. Perhaps alluvial fans migrated into the Belt Sea, or earthquakes on the many active faults along the basin margin triggered underwater landslides and turbid flows of sediment. Regardless, deposition of vast amounts of sediment that now cover 77,000 square miles of the northern Rocky Mountains had begun.

Fragment of Archean gneiss in a coarse sand matrix.

Matrix-supported conglomerate of the LaHood Formation exposed a few miles east of LaHood in the Jefferson River Canyon.

Going-to-the-Sun Road in Glacier National Park

Sediments of the Belt Sea

If you were able to transport yourself 1.5 billion years into the past, Montana would be more like Mars, only with water. Imagine a hot, hostile world devoid of land plants and covered with sand. The Belt Sea, the water body in the Belt Basin, lapped onto this barren landscape, waxing and waning in size as it was fed by rivers and perhaps incursions of ocean water. For 70 million years, the basin subsided and filled to a depth of more than 10 miles with fine-grained sediment called the Belt Supergroup, the thickest pile of sedimentary rocks known on Earth. Although it lacked diverse plants and animals, it teemed with life like sticky mats of cyanobacteria that trapped sediment, making layered structures called stromatolites. Without animals to churn up the sediment, mud cracks, ripple marks, and

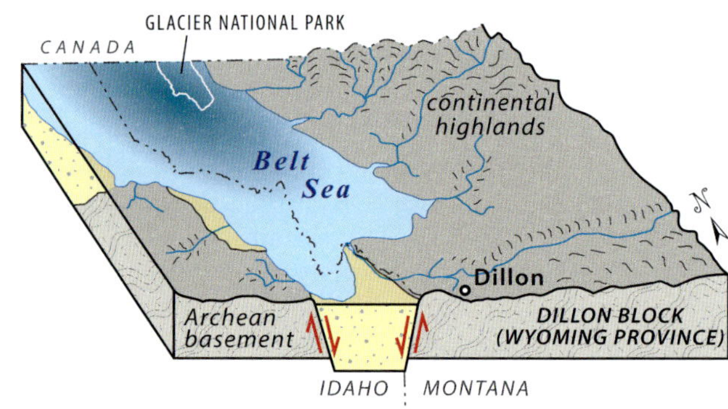

Reconstruction of the Belt Basin during deposition of the Helena and Wallace Formations. —Hyndman and Thomas, 2020

Domal stromatolite in the Wallace Formation. (48.6967, −113.7173)

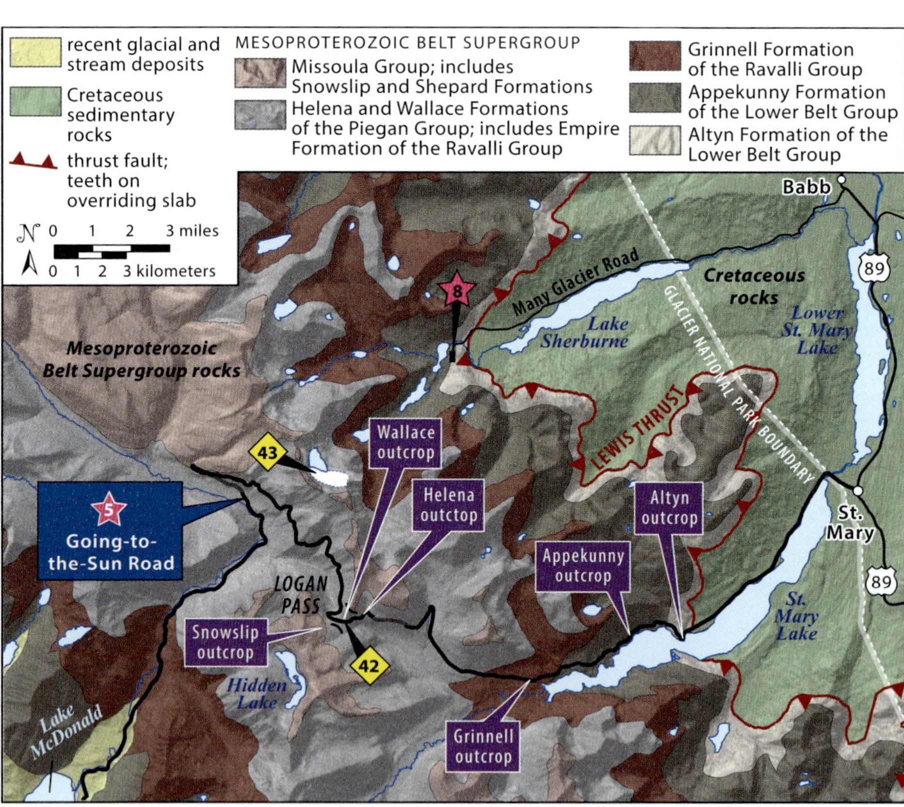

From St. Mary, travel west on the Going-to-the-Sun Road to Logan Pass, stopping at exposures of the Altyn, Appekunny, Grinnell, Helena, and Wallace Formations. Walk the trail to Hidden Lake Overlook to see the Snowslip Formation.

raindrop impressions were exquisitely preserved everywhere in Belt rocks.

One of the best places to see these Mesoproterozoic rocks is on the east side of Logan Pass on the Going-to-the-Sun Road in Glacier National Park. The road transects a pile of Belt rocks above the Lewis thrust, a compressional fault that pushed them eastward in Cretaceous time (site 19). They are folded downward, so the road west to Logan Pass climbs through older to younger rocks of the Altyn, Appekunny, Grinnell, Empire, Helena, Wallace, Snowslip, and Shepard Formations.

Over time, environments changed, and different sediments piled up as the water level rose and fell across the basin. At the bottom of the stack, dolomitic limestone and quartz sand of the Altyn Formation accumulated in nearshore environments where mats of cyanobacteria made stromatolites. The overlying Appekunny Formation contains layers contorted when clouds of coarser sediment dumped into water-saturated, offshore mud. When the water receded, mudflats of the red-colored Grinnell Formation dried out, cracked, and were ripped up into chips during storms, rolled into mudballs, and mixed with quartz sand transported here by streams. When the sea expanded yet again, silty limestone of the Helena Formation and stromatolite-rich limestone of the Wallace Formation were deposited, then covered with cracked and rippled mudflat sediments of the Snowslip Formation when sea level fell. Water rose one last time, depositing fine sand and mud of the Shepard Formation, completing the Belt story in Glacier National Park.

Contorted bedding in the Appekunny Formation. (48.6885, −113.5581)

Mud cracks in the Snowslip Formation. (48.6941, −113.7252)

Sandy, dolomitic limestone in the Altyn Formation. (48.6878, −113.5272)

Armored mudball in the Grinnell Formation. (48.6739, −113.6139)

Silty limestone in the Helena Formation. (48.6988, −113.7051)

6 Kootenai Falls

Ripples in Stone

Mesoproterozoic Belt Supergroup rocks are well known from Glacier National Park, but many special, less crowded places across western Montana feature these spectacular rocks as well. Among the gems is Kootenai Falls, located on the Kootenai River between Libby and Troy. The river plunges a combined 200 feet over two waterfalls and down a spectacular series of steep rapids over horizontal layers of the Mt. Shields Formation, a relatively young Belt formation composed of sandstone, mudstone, and limestone deposited in the Belt Sea some 1.4 billion years ago.

A close look at the rocks shows the telltale signs of deposition in shallow water. The sandy rocks are loaded with ripples, some formed by gentle winds oscillating water and sand into symmetric piles, and others formed by water flowing in one direction, bouncing sand grains into asymmetric piles. See if you can tell the difference in symmetry by running your finger across the ripple crests. The muddy rocks also display polygon-shaped structures formed when clay-rich sediments were exposed, desiccated, and contracted into four-, five-, and six-sided mud cracks, which later filled in with sediment. Passing storms left raindrop impacts so exquisitely preserved that splash rims show the direction of the wind. Within the mud, salt crystals formed that later dissolved and were replaced by mud, preserving the cubic shape of the crystals. Below the nearby suspension bridge, beds of limestone have many layered, mound-shaped structures called stromatolites. They formed in water that was clear and shallow enough that sunlight could penetrate to the bottom, allowing microbial mats of photosynthetic cyanobacteria to flourish.

As the crow flies, Belt rocks progressively thicken westward across the basin and abruptly end near Spokane, Washington. Where did they go? Well, around 750 million years ago, rifting of the western margin of North America during the breakup of the supercontinent of Rodinia lopped off the Belt rocks and carried them elsewhere. Some geologists suspect the detached piece of North America now resides as part of the Siberian landmass, some think it's in Antarctica, while others think the missing Belt rocks are down under in Australia. Well over a century of work has been done on these rocks, and the Belt Basin still retains many unsolved mysteries.

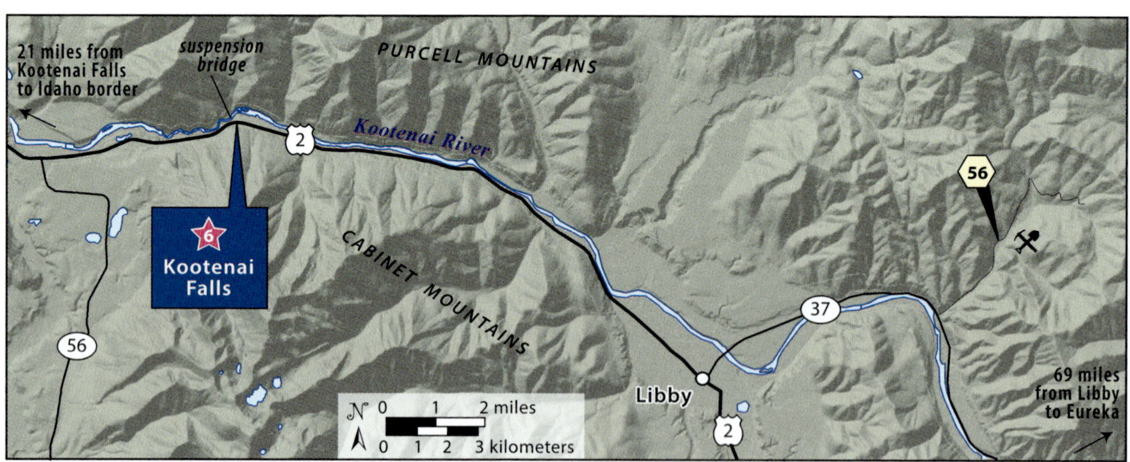

From Libby, travel 11.3 miles west on US 2 to Kootenai Falls Park. Walk the trail east to see the ripple marks at the falls and west to see the stromatolites at the suspension bridge.

Large domal stromatolites formed by cyanobacterial mats in the Belt Sea.

Symmetrical ripples formed by gently oscillating currents moving sand back and forth in the shallows of the Belt Sea during deposition of the Mt. Shields Formation.

Medicine Lodge Pass

Shatter Cones of a Meteorite Impact

Like all celestial bodies in our solar system, Earth gets hit from time to time by very large meteorites and comets. These large impacts are rare events, but when they occur, they can cause global environmental changes that result in the mass extinction of life across the planet, like the famous rock from space that did in the dinosaurs at the end of Cretaceous time. Such an event happened around 900 million years ago, leaving behind evidence the trained eye can see in the rocks around Medicine Lodge Pass on the Big Sheep Creek Backcountry Byway in southwest Montana. Multicellular life was in its infancy at the time, so it's hard to know how the meteorite impacted life on Earth, but given its size, it must have been a calamitous event of planetary scale.

From Dell, travel 2.2 miles southeast on the Westside Frontage Road, and turn right (west) on the Big Sheep Creek Road. Follow the dirt road 31.5 miles to "The Old Bannack Road" sign at Medicine Lodge Pass and the rock exposed west of the sign. (44.6488, −122.9798)

The evidence for the impact is not obvious to anyone passing through the area. There is no large crater to gaze into, but you can find curious looking, conical rocks with striations or grooves called shatter cones. These structures typically form when large meteorites strike Earth, creating high-pressure shock waves that compress and deform the rocks below. In the field, the shatter cones have distinct cone-in-cone striations, and under a microscope, quartz and feldspar grains show fractures and other shock metamorphic features commonly associated with meteorite impacts. The deformed rock is a Mesoproterozoic quartzite called the Gunsight Formation, a Belt Supergroup unit of the Lemhi Group.

Although the shatter cones are now in Montana, the impact didn't happen here but rather near Challis, Idaho. The impact-deformed rocks were shoved here as part of a large thrust slab during Cretaceous mountain building. Determining the size of an impact crater that formed more than 100 miles away and has since been deeply eroded is no easy task, but geophysical data show the crater was about 63 miles in diameter, making it the second-largest known impact structure in the United States, and one of the largest

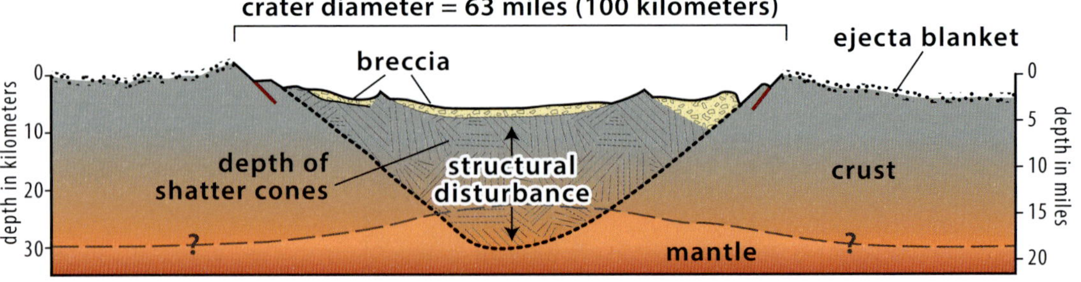

Cross-sectional interpretation of the Beaverhead meteorite impact site shortly after impact. Notice the tremendous depth of formation of the shatter cones.
—Modified from McCafferty, 1995

impact craters on Earth. Depending on its composition, a crater of this size was likely made by a meteorite ranging from 4 to 10 miles in diameter. Although not seen in the rocks at Medicine Lodge Pass, the Gunsight Formation also contains randomly distributed, small breccia dikes and veins of a glassy rock that can form by melting during meteorite impacts. Isotopic dating of minerals formed by recrystallization of this impact glass is how geologists determined that it likely happened around 900 million years ago.

The processes that make shatter cones suggest that they formed miles below ground at the time of the impact, so a lot of overlying rock has since been removed to expose the depths below the site of the impact. These rare structures are something special, so please don't hammer on them or take samples. They are of greater educational value in the field than in a rock garden.

A prominent shatter cone exposed at Medicine Lodge Pass on the Big Sheep Creek Backcountry Byway. Notice the striations or grooves along the side of the cone.

Purcell Sill from Many Glacier Hotel

Magma Injected during the Breakup of Rodinia

Following deposition of the Belt rocks, there is no evidence for what happened in Montana for the next 600 million years. The absence of a rock record reflects a period of tectonic inactivity and erosion of older topography. Such periods of erosion create unconformities, gaps in the rock record that can be as informative as the record itself. The inactivity ended with an event of epic proportions, the breakup of the supercontinent called Rodinia. In Montana, the evidence of the breakup comes from dark-colored igneous intrusions, the most notable being the Purcell Sill in Glacier National Park.

Recall that the Mesoproterozoic Belt Supergroup was deposited in an inland sea formed as the Nuna (Columbia) supercontinent began to break apart. Rodinia formed around 1.2 to 1.0 billion years ago through the assembly of continental fragments from the breakup. Supercontinents don't last long because they act like blankets on the mantle, trapping and building heat that ultimately tears them apart. This happened to Rodina around 750 to 633 million years ago. One of the early signs of the pending demise of a supercontinent is the injection of dark-colored igneous rocks because the stretching and release of pressure melts parts of the mantle. The Purcell Sill is one such intrusion, having injected between and parallel to limestone layers of the Wallace Formation in the park area around 780 million years ago. In places, this amazing sheet of magmatic rock is up to 100 feet thick. It ramps and evenly splits into two sills, but it mostly stays in the Wallace Formation over an area of 50 square miles.

The resulting black stripe in the contrasting gray limestone demands attention and adds a special beauty to places like Mt. Gould as seen from Many Glacier Hotel. Even from this distance, the black sill and the layers of altered limestone above and below it stand out. When the sill injected, it was about 2,000°F, so the Wallace limestone, composed mostly of calcite, was cooked. The heat baked out any dark organic matter, bleaching it white above and below the sill and

From Babb, travel 12 miles west on the Many Glacier Road to Many Glacier Hotel. The Purcell Sill is visible in the distance to the southwest on Mt. Gould.

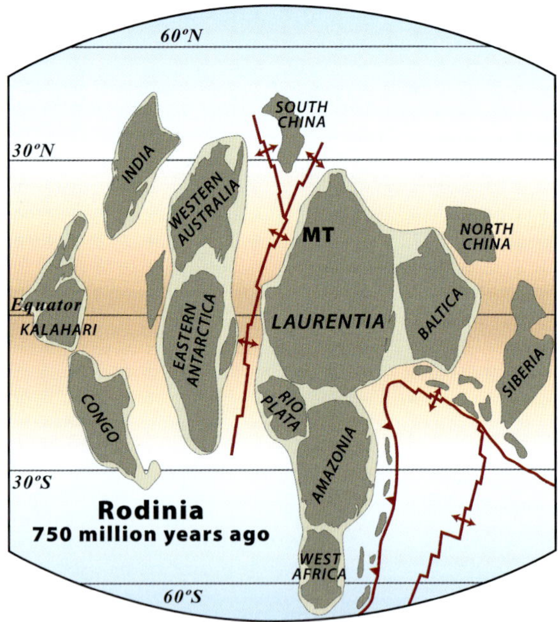

Tectonic extension and breakup of the supercontinent of Rodinia around 750 million years ago injected basaltic diabase dikes and sills into the Belt Supergroup and older rocks in Montana (western Laurentia). —Modified from Torskik and others, 2008

The Purcell Sill exposed on Mt. Gould as seen from Many Glacier Hotel. The basaltic diabase sill is the prominent, dark layer in the Wallace Formation limestone.

A close-up of the basaltic diabase sill and white marble on the Going-to-the-Sun Road. The roadcut is at 48.6969, −113.7114.

enlarging the crystals to make a metamorphic rock called marble. Along the contact with the cold limestone, the sill cooled more quickly, so it is finer grained than in the middle of the sill where it cooled more slowly. On close inspection, it can be hard to see crystals because they didn't grow very large due to rapid cooling. In other places, white-colored crystals of plagioclase feldspar and dark-colored crystals of pyroxene, olivine, and hornblende are clear to see. Such minerals grouped together are called basaltic diabase (dark) and diorite (light). A good place to see these rocks up close is next to the Going-to-the-Sun Road just below (east of) Logan Pass.

Camp Creek near Melrose

The Great Unconformity and a Rising Cambrian Sea

Bob Dylan once wrote, "how many years can a mountain exist, before it is washed to the sea?" The answer is recorded in the rocks at Camp Creek near Melrose, Montana. At this special place, 520-million-year-old Cambrian beach sandstone rests on the eroded guts of Paleoproterozoic mountains, igneous and metamorphic rocks formed about 1.8 billion years ago. The old rocks were smashed, deeply buried, and heated during the collision of continents of the Big Sky mountain building. Notice there is a dike of igneous rock with big crystals—a pegmatite—that cuts the metamorphic layering. It's a squirt of magma formed when rocks melted deep inside Himalayan-sized mountains.

As gravity, water, and ice were busy tearing the mountains down, its deep core raised them back up, much like a ship rising in the water while being unloaded of goods. After 1.3 billion years of battle, the core was gone, and the mountains eroded to sea level. After Rodinia split apart along spreading ridges, the planet was extremely cold and covered in ice several times between 717 and 635 million years, the so-called Snowball Earth of the Cryogenian Period. Complex, multicellular life emerged as the planet warmed during the Ediacaran Period. As the ice melted and the new ocean spreading ridges grew wider and higher, ocean water rose and slowly flooded the eroded surface, burying it in sand and creating the first beach in Montana in nearly 1 billion years. The huge gap in time between the gneiss and the beach sand is known as the Great Unconformity. The erosional surface was remarkably flat, like it is over much of the continent.

This rising sea is the start of the Cambrian transgression, a marine incursion that marks the beginning of the accumulation of thousands of feet of sedimentary rock on the continent over the next 275 million years. Sea levels rose and fell, while the land subsided from the weight of that sediment to accommodate the growing pile. We know the rate at which the rising ocean flooded North America was very slow because the oldest beach sand here is nearly 20 million years older than the same deposits more than 600 miles to the east in North Dakota. Montana was located just south of the equator, so the rising, well-oxygenated water was warm and supportive of tropical life, which included all the major groups of multicellular animals and numerous marine plants. Crab-like animals called trilobites left behind countless shells as they shed them to grow. The correlation of trilobites from one place to another enabled geologists to figure out that

Worm feeding burrow (the linear, tube shape parallel to the top edge of the sample) in the muddy shelf deposits of the Wolsey Shale.

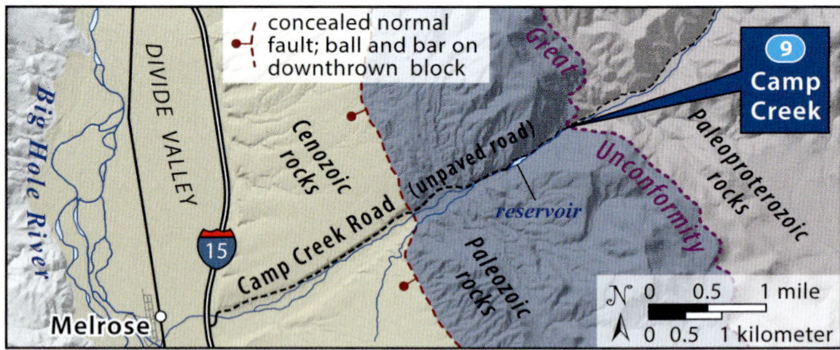

From the I-15 Frontage Road at Melrose, travel 4 miles northeast on the Camp Creek Road to the pullout on your left (north), and walk along the rock wall to observe the Great Unconformity. Walk west (upsection) along the Camp Creek Road to see the rocks deposited as sea levels rose up to cover part of the continent.

The Great Unconformity at Camp Creek east of Melrose. The geologist's hand is on the Great Unconformity, a gap in the rock record spanning 1.3 billion years at this location.

offshore carbonate sediment accumulated in Montana while quartz sand was washing back and forth on coastal beaches in the Dakotas.

The pegmatite dike ends at the Great Unconformity, where it was eroded and incorporated into the lowermost deposits of the overlying Cambrian Flathead Sandstone. The chunks are angular at the base because they were not moved very far from their source, some resting in scour troughs eroded into the underlying metamorphic rocks. The sand consists of well-rounded and uniformly sized grains of quartz deposited in parallel layers, typical of sand rolling back and forth in the surf of a beach. A bit higher up in the pile, the sand forms mounds or hummocks interbedded with shale. This is typical of storm deposition in the submerged portion of a beach, showing that the ocean was rising and deepening on its trek eastward. Fossils of brachiopods, trilobites, and worm burrows in shale beds show us the continent was on its way to becoming a tropical waterworld. As the sea went in and out, the land subsided under the weight of the accumulating sediment, and it stacked up into a multilayered pile, oldest rocks at the bottom.

Walking westward on the Camp Creek Road takes you through about 30 million years of shallow, tropical marine, sedimentary rocks deposited in Cambrian time. We start the journey at the bottom of the stack and move upwards: the Flathead Sandstone; then shelf mud of the Wolsley Shale; then, about 1 million years later, offshore limey sand and mud of the Meagher Limestone; then sand, mud, and limestone of the Park Shale and Pilgrim Limestone. The deposition of each unit ended when the sea level dropped, exposing and eroding the shelf before the next incursion of ocean water. An influx of quartz sand near the top of the Pilgrim Limestone tells us that land sources were emerging as the sea level dropped, leading to exposure and erosion of the shelf prior to the next sea-level rise that buried it in sand and mud of the Red Lion Formation.

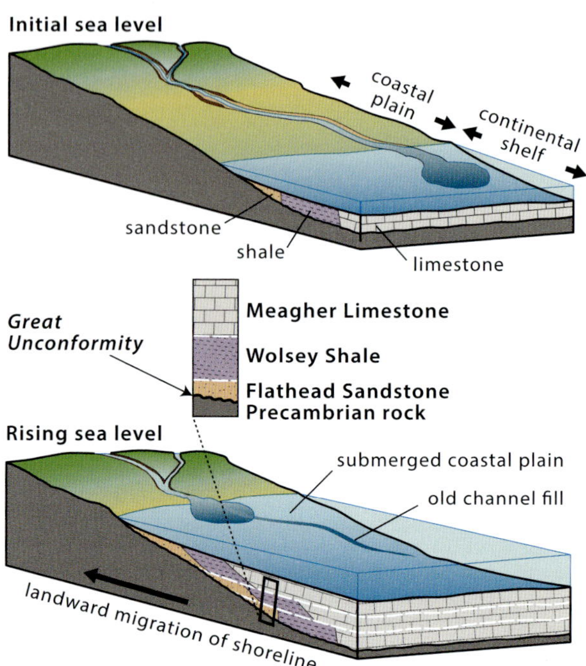

Sequence of rock types and formations deposited as sea levels rose during the Cambrian.

The shell of a trilobite, a common animal living in the Cambrian oceans.

Oncolites in the Meagher Limestone are thinly layered balls of limestone made by encrusting microbes living in the shallow, tropical waters of the offshore carbonate platform.

Beach deposits of the Cambrian Flathead Sandstone lie directly above the Great Unconformity. It shows an upward transition from parallel layered sand of the upper beach to shale and wave-caused, sandy hummocks deposited when the sea level rose.

One possible cause of sea-level drop is global cooling. As ice builds up on land, water is removed from the oceans, and the sea level drops. Each time Cambrian oceans receded, warm water animals were decimated, while those living in deep, cold water survived. Perhaps removal of the greenhouse gas carbon dioxide to make limestone cooled the atmosphere and the oceans. For reasons not well understood, uppermost Cambrian, Ordovician, and Silurian rocks were eroded away here, and the erosional surface is buried by dark-colored, sandy dolostone of the Devonian-aged Jefferson Formation, deposited during another big rise in the sea across the North American continent.

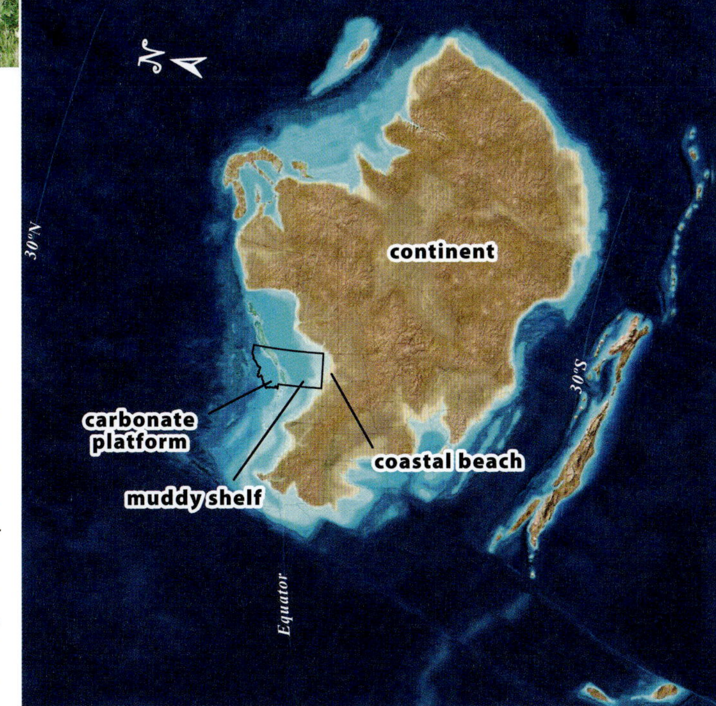

Paleogeographic map of the Cambrian transgression of the continent around 500 million years ago. The north-northwest-trending feature in Montana is a carbonate shoal. —Image © 2023 Colorado Plateau Systems

10 Bighorn Canyon

A Grand Cut through Paleozoic Rocks

Geologists love the Grand Canyon in Arizona, in part because of its thick pile of undeformed Paleozoic sedimentary rocks. Similar-aged rocks in Montana's mountains tend to be buggered up by compression, intruded by granitic rocks, and torn apart by extensional faults. East of the mountains, tectonic influences dissipate but so do exposures of Paleozoic rocks. Luckily for geologists, some are found gently folded in ranges uplifted by Laramide compressional faults, like in the Pryor and Bighorn Mountains south of Hardin. The Bighorn River passes through these mountains, eroding into the rising land and exposing up to 1,500 feet of mostly undisturbed Paleozoic sedimentary rocks in the steep walls of one of the grandest canyons in Montana.

There are two access points to the Bighorn Canyon National Recreation Area, Highway 313 south of Hardin in the north and Highway 37 from Lovell, Wyoming, in the south. This book focuses on overlooks accessible from the southern entrance. Highway 37 follows close to the rim of the canyon with unparalleled views of the steep walls dominated by the thick Mississippian Madison Group limestone. This tropical marine limestone is loaded with fossil corals, crinoids, and brachiopods that lived in the warm ocean waters that covered Montana from 359 to 326 million years ago. The best views are from Devil's Canyon Overlook and Sullivan's Knob Trail, where you can make echoes against the canyon walls.

The upper part of the Madison Group here has a distinctive zone where water dissolved the limestone, creating karst features like caves and breccias where sinkholes collapsed. The dissolution likely formed during a period of sea-level drop and exposure of the limestone to rainwater, which is slightly acidic. Some geologists argue that the red color of the overlying Amsden Group is from soil formed during millions of years of weathering and dissolution of the exposed Madison Group limestone. The weathering concentrated the insoluble, oxidized (red) clays. Perhaps the red color comes from oxidation of iron-bearing sediment deposited during Amsden time or staining of the sediments by iron-rich groundwater long after its deposition. In any case, surface runoff from the Amsden trickles down the light-colored cliff walls of Madison Group limestone, leaving behind red streaks as it stains the rock.

BIGHORN CANYON
SOUTHWEST
NORTHEAST
Bighorn River
Paleozoic rocks
thrust fault
Archean basement
not to scale

Cross section of Bighorn Canyon showing the Laramide folds and faults. —Modified from Hyndman and Thomas, 2020

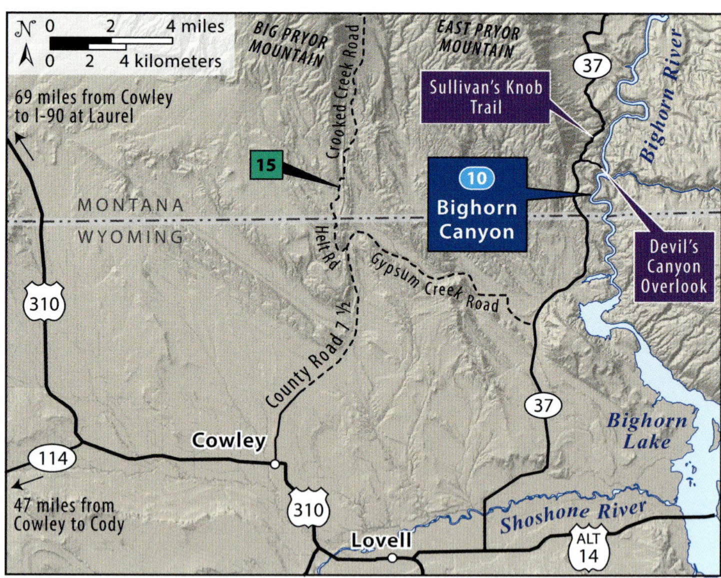

From Lovell, Wyoming, travel 3 miles east on US 14 to WY 37. Turn left (north), and travel 15.6 miles north to Devil's Canyon Overlook. Sullivan's Knob trailhead is 0.9 miles north, with an easy 0.75-mile round trip walk on a good trail to the overlook. Note the Bighorn River's entrenched, meandering path north of Bighorn Lake.

Although the folds and faults of the Pryor and Bighorn Mountains formed during Laramide compression, the Bighorn River canyon is much younger. Notice that the river cuts across the folds and faults and has the sinuous shape of a low-gradient, meandering stream. It once flowed over a flat surface, likely in loose sediment that covered the folded rock. The plains have recently been on the rise due to crustal stretching in the western United States, so as the region was slowly uplifted, the sinuous river cut downward into the hard rock below, becoming trapped in the fossil meanders of its old channel.

Brachiopod fossil in the Madison Group limestone.

The Bighorn Canyon viewed looking south from the overlook on Sullivan's Knob Trail.

Gates of the Mountains

Fossils in the Madison Group Limestone

Since people have lived in Montana, rocks have been important landmarks and none more so than those formed in the Mississippian Madison Group limestone. These marine deposits are composed of calcite, a mineral that tends to dissolve in humid climates where carbon dioxide from decaying organic matter mixes with water to form carbonic acid. Montana has a dry climate, however, so limestone resists erosion, forming prominent exposures like Beaverhead Rock near Dillon and steep canyon walls like in the Bighorn Canyon (site 10). Choosing one place to show off the Madison limestone is no easy task, but boat rides are fun, so let's explore the Gates of the Mountains near Helena, Montana, just like Lewis and Clark did in the summer of 1805.

The walls of the canyon consist of folded Mississippian Madison Group limestone, deposited when Montana straddled the equator and was covered in tropical marine water. From 359 to 326 million years ago, shells of rugose corals, tabulate corals, crinoids, brachiopods, bryozoans, and other calcite-secreting animals accumulated on the seafloor to form limestone. The water was very clear, too far from land to receive suspended sediment, which makes it difficult for animals to filter the water for food or for marine plants to photosynthesize. As the ocean began rising, mud and shell-rich limestone of the Madison Group's Lodgepole Formation accumulated on a quiet, tropical shelf seldom stirred up by storm waves and currents. Over time, the shell pile grew and spread, forming a huge carbonate platform of the overlying Mission Canyon Formation, also of the Madison Group. It was not a reef, however, because poorly oxygenated ocean water and global cooling had decimated reefs in Devonian time.

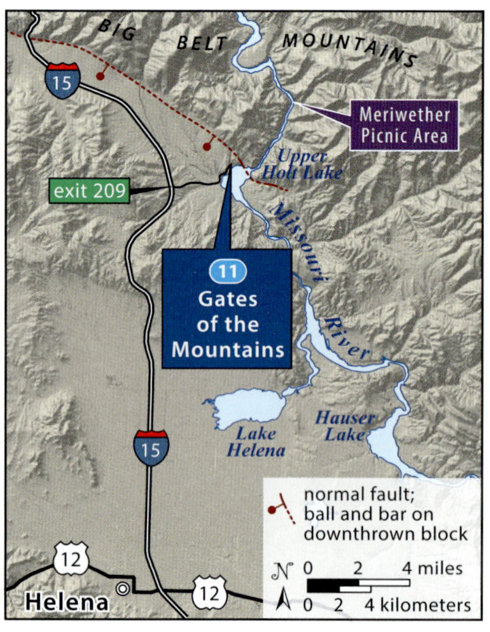

From Helena, travel 16 miles north on I-15. At exit 209, turn right (east) on Gates of the Mountains Road, and travel 2.8 miles to the parking lot for the boat tour.

The Missouri River has cut deep into the Mission Canyon Formation, which forms the steep canyon walls. It seems logical that the Missouri River should go around the Big Belt Mountains, but instead, it turns to the northeast and goes through them. Captain Lewis noted in 1805 that it appeared as if the river "forced its way through this immence [sic] body of solid rock." To cut this path, the river must have been flowing here before the mountains were uplifted. Crustal stretching began raising the northern end of the Big Belt Mountains along a normal fault and dropping down the Helena Valley around 17 million years ago. As the mountains went up, the river eroded into the hard limestone and then became trapped in its canyon.

During the summer months, a charter boat provides regular trips through the canyon past the towering cliffs of the Madison Group. The limestone is intensely folded, a product of compression during the Sevier mountain building in Cretaceous time. Caves in the limestone were dissolved by slightly acidic groundwater and exposed when the river cut the canyon. You can get off the tour boat at the Meriwether Picnic Area to look closely at the rock, which is loaded with fossils. Please do not collect them because they should remain for everybody to see. If you want to collect Mississippian fossils like these, consider the spillway at Clark Canyon Reservoir south of Dillon (45.0015, −112.8612). It's a great place to take kids because it's accessible, and whole fossils weather out of the rock and can easily be picked up by small hands.

Artist rendition of the tropical marine paradise in Montana about 360 million years ago. —Courtesy of Douglas Henderson

Typical fossils in the Mississippian Madison Group limestone. 1. Brachiopods; 2. Rugose corals; 3. Crinoids.

View looking upstream into the Gates of the Mountains.

12 Bear Gulch

A Fossil Smorgasbord in a Mississippian Lagerstätte

From the title, you might be thinking, what is a lagerstätte, and can I get one at my local pub? It's a sedimentary deposit containing extraordinary fossils with exceptional preservation, typically including the details of soft tissue. Such fossil occurrences, while scattered throughout the geologic record, are extremely rare. A few of the more famous examples include the Cambrian Burgess Shale in Canada, the Eocene Green River Formation in Wyoming, and the Jurassic Solnhofen Limestone in Bavaria, famous for the feathered theropod dinosaur *Archaeopteryx*. We are quite fortunate to have such a deposit in Montana, the less famous but equally spectacular Late Mississippian Bear Gulch Limestone between Grass Range and Lewistown. The site is on private land, but summer tours of the outcrops are available through BearGulch.net, and many excellent fossils are curated into the Melton Collection at the University of Montana Paleontology Center in Missoula.

The Bear Gulch Limestone is a lens of the Heath Formation of the Big Snowy Group, deposited about 324 million years ago when the sea rose to bury the karst surface of the Madison Group in sand, mud, and limestone. The fossils are preserved in a very fine-grained, limey sediment that accumulated in quiet water, perhaps a protected embayment where organisms were buried so rapidly that they could not be destroyed by scavengers or broken down by bacterial decay. The fossils are dispersed throughout the lens, so they did not perish in one event. Perhaps the water was not salty enough, too salty, or too low in oxygen to allow scavengers access to the soft tissue, an idea supported by fossils of bottom-dwelling fish with distended gills, which happens when they die from asphyxiation.

The Bear Gulch Limestone is one of the world's most diverse and well-preserved fossil fish collections, with more than 3,000 specimens representing at least 150 species. Common groups include bony fish, sharks and rays, and enigmatic swimming animals, like the conodont-eating mollusk *Typhloesus*. Marine plants and invertebrate fossils include algae, sponges, clams, snails, squids, brachiopods, bryozoans, starfish, horseshoe crabs, and worms. The vertebrate fossils are typically complete specimens with preserved internal organs, eyes, and even skin pigments that show color patterns. Scientists learned about the biology and behaviors of these animals because the fossils are preserved so well, including food interests from gut contents and love interests from animals fossilized while mating. Age data suggest the lagerstätte represents only 1,000 years of sediment accumulation, an instant in geologic time.

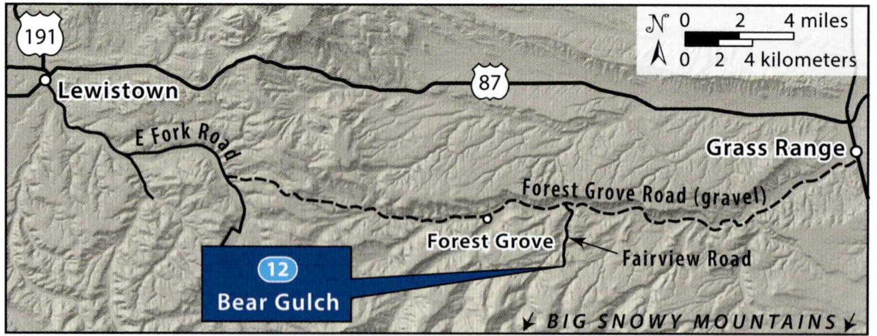

From Grass Range, travel 12.6 miles west on Forest Grove Road, turn left (south) on Fairview Road, and travel 2.5 miles to the Bear Gulch Pictographs. The private site is about 25 miles east of Lewistown. For reservations for summer tours: https://beargulch.net

Fossil of a 5-inch-long cartilaginous fish from the Late Mississippian Bear Gulch Lagerstätte. This specimen is on display at the Museum of the Rockies in Bozeman. —James St. John, Wikimedia Commons CC 2.0

13 Dalys Spur

Dune and Shelf Sand on Pangea's Coast

For decades, geology students have learned to measure and describe sedimentary rocks at Dalys Spur, south of Dillon. This impressive rock looms over the Beaverhead River, exposing the Quadrant, Phosphoria, and Dinwoody Formations, which are tilted to the west about 50 degrees. Measuring a section of rocks is done with a 1.5-meter-long Jacob's staff, the section sketched, and data recorded in a notebook. The goal is to use sedimentological and paleontological attributes of the sedimentary rocks to identify the likely environments of deposition. Sedimentary structures are crucial because they tell us how the grains moved. For example, sand rolls back and forth on a beach, coming to rest in parallel layers, while sand on a dune hops, creating large cross-beds that are at an angle to the overall bedding.

Stand back and look at the Pennsylvanian-aged Quadrant Formation. The internal sandstone layers are at angles to the bedding and are up to 10 feet thick. This large-scale cross-bedding forms in migrating dunes where sand piles up. Older Paleozoic rocks in Montana were deposited in clear, tropical oceans, but by Pennsylvanian time, regional uplift caused the retreat of the seawater. Pangea, the last of the supercontinents, was assembled, and arid, coastal sand flats and dune fields formed along its western margin. Notice the glassy grains of quartz in the sandstone are all the same size

Erosion by the Beaverhead River exposed the Quadrant, Phosphoria, and Dinwoody Formations at Dalys Spur, south of Dillon.

with rounded edges, typical of grains moved by the wind. These grains came from the continent to the east and were blown into large coastal dunes.

The overlying Phosphoria Formation is sandy, but the grains are not uniform, and the bedding looks churned up, as though animals moved through it as they searched for food. These shelf deposits, including chert (formed from sponges) and phosphatic black shale (formed from the bones of fish that flourished in an upwelling zone), were deposited as the sea level rose in Permian time.

As the sea level dropped again at the end of the Permian Period, 252 million years ago, life on Earth was nearly wiped out, with the loss of about 90 percent of all species. The cause of this extinction event, called the Great Dying, is debated, but changes in climate and ocean chemistry caused by all the continents being in one place on Earth's surface, along with volcanic eruptions above a plume of heat rising in the mantle beneath Siberia, are likely culprits. When the sea returned in Triassic time, shelf deposition returned with the Dinwoody Formation, the last of the great western margin ocean deposits in Montana. It contains brachiopod species that survived the extinction. Storm waves and currents moved them into piles, and as the storm waned, silt and clay settled from the water, burying the shells, creating graded beds. The shelf must have been shallow and prone to storms because the limestone-to-siltstone sequence of graded beds are repeated multiple times in the Dinwoody. Dalys Spur is sacred to the First People, so please be respectful, and do not use hammers or take fossils.

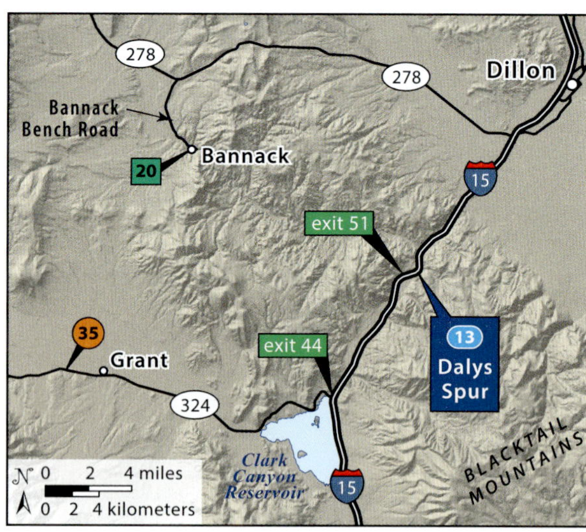

From Dillon, travel 11.2 miles south on I-15 to the Dalys turnoff (exit 51). Make a U-turn, and travel northeast on the old frontage road to the prominent rock spur, and walk west (upsection) across the Pennsylvanian through Triassic rock exposures.

Large-scale cross-bedding in the Quadrant Formation was produced by a migrating coastal sand dune.

An unknown, small (half inch long) fish fossil in phosphate-rich (white sand grains) shale of the Phosphoria Formation.

Lingulid brachiopod shells in limestone of the Dinwoody Formation.

A vertical burrow, about 6 inches in diameter, cuts across tilted layers of shelf sand. It may have once housed a crab-like animal, but no body fossils have ever been found within these burrows.

Block Mountain

A Sedimentary Basin Overrun by Sevier Thrusting

Like all supercontinents, Pangea was short lived and started breaking up around 200 million years ago, at the beginning of Jurassic time. The Atlantic Ocean opened, and the new North American plate moved westward, colliding with ocean plates and volcanic islands to the west. The denser oceanic plates subducted into the mantle off the continental coastline, generating magma and volcanoes. Farther inland, slabs of sedimentary rock moved eastward on thrust faults, creating mountains in western Montana called the Sevier fold-and-thrust belt. The thick pile of stacked slabs added weight to the crust, bending it down east of the mountains and forming a lowland, called a foreland basin, parallel to the front.

Oblique aerial view of folded and thrust-faulted sedimentary rocks at Block Mountain. The Sandy Hollow thrust is a spectacular fault along which anticlines and synclines were shoved up and over other folds. —Base image courtesy of Google Earth

From Dillon, travel 17.5 miles north on MT 91 to the Burma Road south of Glen. Turn right (east), and travel 7.4 miles to Sandy Hollow Gulch Road, and turn left (north). Park at the prominent "Y" in the dirt road, and walk across the steeply tilted Mesozoic sedimentary rocks exposed in the east half of the Sandy Hollow anticline.

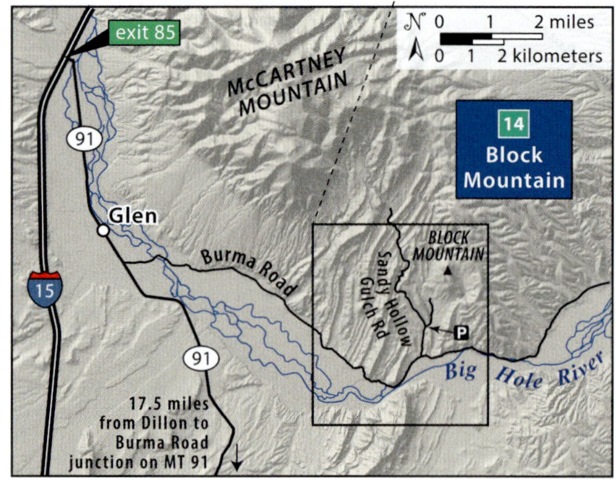

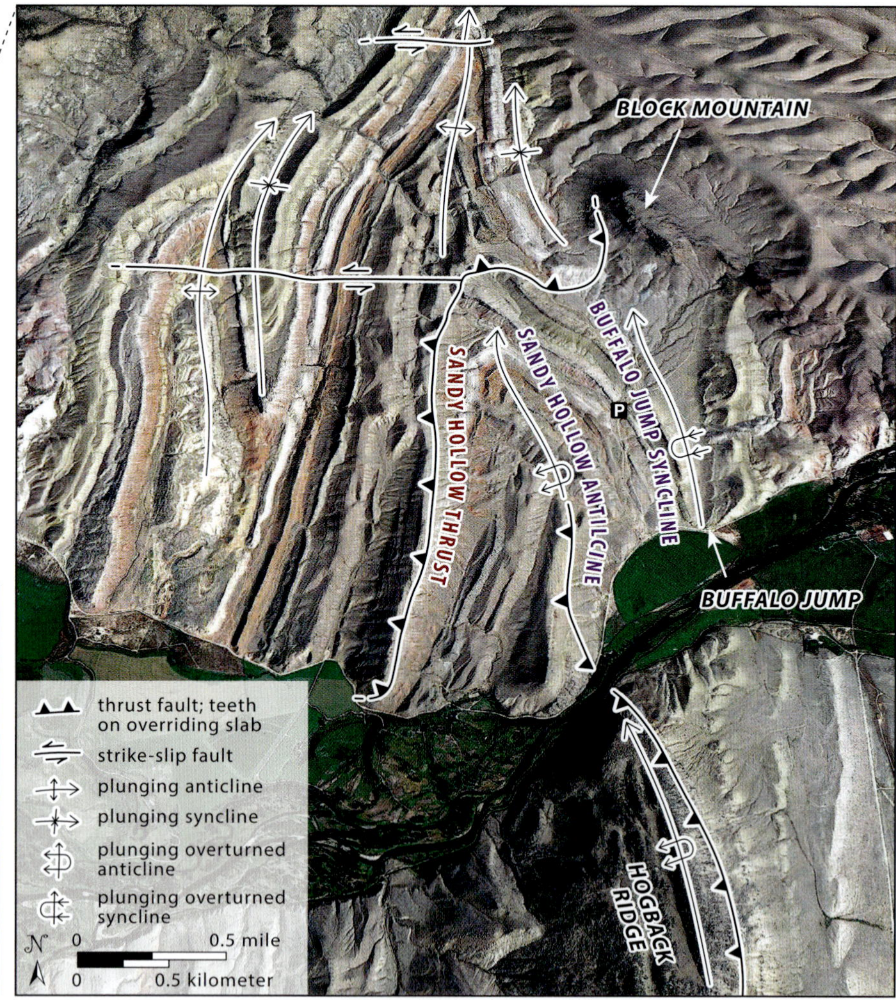

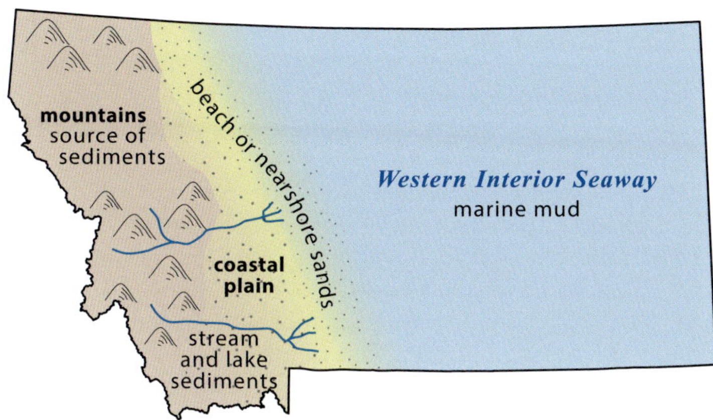

Cretaceous paleogeography and depositional environments in Montana. —From Hyndman and Thomas, 2020

A fossil snail from lakebeds in the Kootenai Formation.

The basin filled with sediment from the eroding mountains, and ocean water flooded in from the Arctic Ocean and the Gulf of Mexico, creating the Western Interior Seaway by the Late Cretaceous.

At Block Mountain north of Dillon, mud and sand of the Jurassic Morrison Formation were shed off the highlands and deposited by streams into coastal lowlands. This formation is famous for its abundant dinosaur fossils, but few had been found here until a distracted geology student discovered a bone while mapping. Some additional exploration resulted in the discovery of a dozen juvenile sauropods—Littlefoot to those who learned paleontology from the film franchise *The Land Before Time*. Although work remains to be done, it appears the dinosaurs drowned while attempting to cross a swollen river about 150 million years ago.

By Cretaceous time, tectonic activity to the west increased the gradient of the land and moved the Sevier mountain front into Montana. A conglomerate at the base of the Kootenai Formation shows that streams had steep enough gradients and water flowing fast enough to transport pebbles and cobbles, not just mud as before. Black chert and white quartzite gravel show us that Paleozoic rocks were now exposed in the highlands and eroding. As streams migrated laterally, channel gravels were buried by point bar sands and eventually

Fossil humerus or arm bone from a hadrosaur, a plant-eating, duck-billed dinosaur that roamed the many streams draining the mountains into the foreland basin.

floodplain muds. Some channels filled with fragments of brightly colored floodplain soil that fell in when cutbanks were eroded. Fossils show us the coastal plain was covered in palms, conifers, and ferns that fed plant-eating dinosaurs, like hadrosaurs.

Two limestone members in the Kootenai Formation show the foreland basin was occupied by lakes at times in the Early Cretaceous. These odd, limy deposits are loaded with snails and clams, some forming beds up to 5 feet thick of unbroken snail shells. Such an environment must have been calm, with gentle waves and currents that piled up the shells without breaking them. The lakebeds must have been loaded with

Stream-deposited conglomerate and sandstone of the Cretaceous Kootenai Formation.

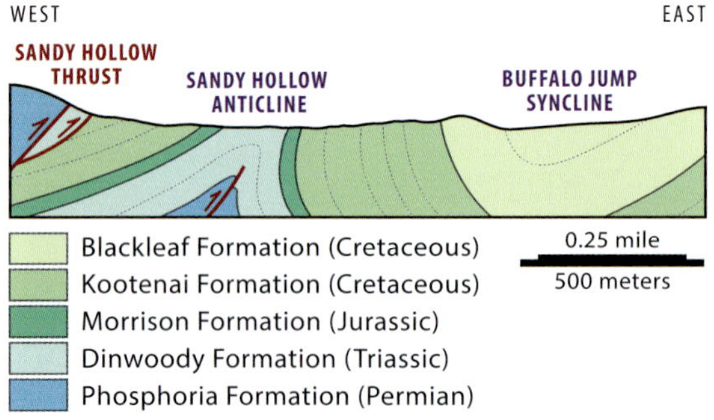

Simplified cross section showing the Sandy Hollow thrust and anticline.

View north into the "rat's nest" where older-over-younger rocks and intense folding and faulting mark where the Sandy Hollow thrust fault ramps upward.

organic matter that was attractive to snails but not good for animals that ate snails. Perhaps they were stagnant lakes formed on the coastal plain during times of tectonic inactivity. Rare shark teeth suggest a connection at times to the interior seaway. Rivers returned to the area with the deposition of delta sand and mud of the Blackleaf Formation. The sand was then buried by black shale when oxygen-poor water from the interior seaway flooded far to the west around 100 million years ago. By the close of Mesozoic time, the eastward migrating Sevier mountain building caught up with these rocks, and they too were folded and thrust faulted.

The Sevier mountain building describes a time of crustal shortening that mostly involved sedimentary rocks that lay above the hard, crystalline basement. Low-angle thrust faults followed weak bedding planes, typically in slippery shales, and thick packages of rocks were shoved from west to east along the faults. Where they could not move forward, faults broke upward, carrying older rocks over younger rocks and in places forming folds where sedimentary rocks draped over the tips of rising fault planes. The fold-and-thrust belt propagated eastward over time, forming a wedge of deformed rock that thins toward the Rocky Mountain front and largely ends where resistant thrust slabs of Belt Supergroup and Paleozoic limestone rise from the plains.

At Block Mountain north of Dillon, the folds and thrust faults are so extraordinary they draw geology field camps from all over the world. The area boasts anticlines (upfolds) and synclines (downfolds) at all scales. The folds are tilted (plunging) north, so erosion has exposed their textbook curvature. Where the fault is in the core of a fold, it likely moved the fold eastward, like raising the edge of your hand through a pile of papers. The Sandy Hollow thrust fault at Block Mountain carried two big folds over other folds. Where it ramps up, the geology is so difficult to map the students call it "the rat's nest." To the east, rising basement blocks stopped forward progress and bent the fault, while other faults moved toward the compression (west), like a car slowly colliding with a wall.

As the fold-and-thrust belt propagated from west to east, it deformed the deposits shed from it into the foreland basin. Some gravels shed off the mountains were even overrun by thrusts, like the Beaverhead conglomerate (site 20). Many thrust faults and folds are intruded by granitic rocks, showing that magma migrated eastward from western sources, following the porous and permeable thrust faults. These thrust-emplaced intrusions occur at McCartney Mountain and in the Pioneer Mountains to the west. Block Mountain is a younger basalt that flowed over the folds in Eocene time. The hard rock resists erosion, so it has become the topographic high point during much younger crustal stretching—the Basin and Range extension—over the last 17 million years.

The plunging anticline in the foreground, part of the Sandy Hollow anticline, formed over a rising thrust fault. On the skyline, Block Mountain is an Eocene basalt that flowed down a river channel eroded into the underlying folded rocks.

15 Vermillion Valley in the Pryor Mountains

Arid Climates and Shallow Seas

While Triassic through Cretaceous deposits in southwestern Montana (sites 13 and 14) show the conditions of North America's western margin as it collided with tectonic plates to its west, deposits farther east reflect more inland environments. The best place to see these rocks is in the Pryor Mountains south of Billings, where you can walk across Triassic through Cretaceous rocks in the Vermillion Valley, named for the red color of the Triassic Chugwater Formation. These sandstones and shales are brick red due to very small quantities of iron oxide coating the quartz grains. When and how it got there is uncertain, but red sediments occurred globally at this time due to the arid conditions across Pangea. These sediments, which were deposited mostly by flowing streams, look so different from formations of the same age to the west that they have different names. A 70-million-year gap in the rock record, an unconformity, exists between the Chugwater and the overlying Jurassic Piper Formation (Ellis Group), perhaps caused by uplift of the land within the foreland basin during the start of the Sevier mountain building.

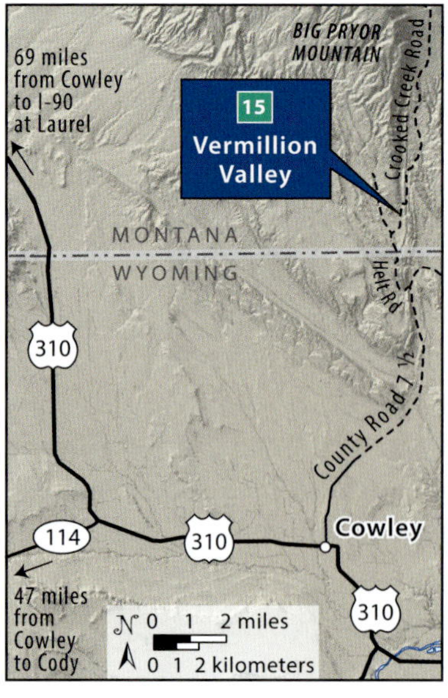

From Main Street in Cowley, Wyoming, travel 5.3 miles north on County Road 7½, and continue straight for another 2.9 miles. Turn left (west) on Helt-Gypsum Springs Road, and travel 1.8 miles, staying to the right for 1.0 mile on Crooked Creek Road. Stop at the pass on the red-colored, Triassic Chugwater Formation for an overview (45.0167, −108.4250). Walk east to cross the steeply tilted Triassic, Jurassic, and Cretaceous sedimentary rocks.

The rocks in the Vermillion Valley are tilted to the east on one side of a syncline forming the valley between the anticlines of the Big Pryor and East Pryor Mountains. Walking east across the syncline takes you through the Chugwater, across the unconformity, and into the lighter-colored Piper Formation. Its shale, limestone, and evaporite deposits (gypsum) were deposited first in lakes, then in an ocean that invaded the foreland basin for the first time, called the Sundance Seaway. The limestone at the crest of the ridge has a few snail and clam fossils and marks a significant incursion of warm, ocean water into the basin. The overlying Rierdon and Swift Formations, both also of the Ellis Group, record continued marine deposition as the Sundance Sea retreated to the north during Middle and Late Jurassic time. They contain abundant fossils of oysters and cone-shaped squids called belemnites. The Ellis Group is missing from southwestern and central Montana, likely due to uplift within the foreland basin.

As the ocean retreated, sediments from the rising Sevier highland in the west built a coastal plain of streams, lakes, and swamps of the Morrison Formation. Sauropod dinosaurs, like those found at Block Mountain (site 14) roamed this place, supported by plentiful vegetation. As the mountains grew in the Early Cretaceous, sandy and gravelly streams of

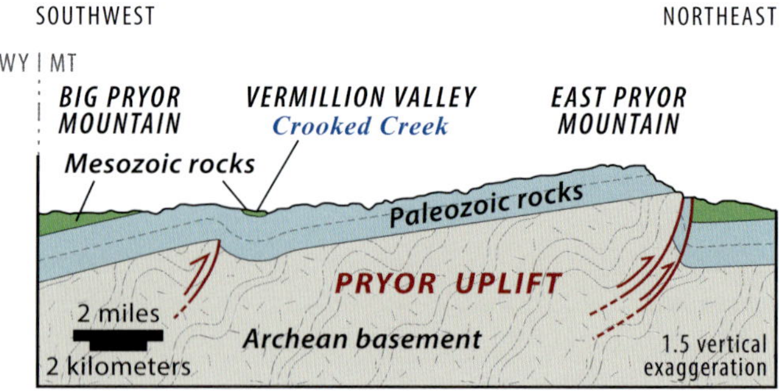

Cross section showing the high-angle faults that formed the tilted sequence of sedimentary rocks exposed along the Crooked Creek Road in the Pryor Mountains. —Hyndman and Thomas, 2020

View of the Mesozoic sedimentary rock layers from the Vermillion Valley Overlook in the Pryor Mountains south of Billings. East Pryor Mountain (in the distance at right) and Big Pryor Mountain (at left) are capped with the Madison Group limestone.

Beds of oyster shells form the resistant ridge of the Swift Formation to the east of the road.

the Kootenai Formation migrated from west to east across the foreland basin, looking much the same over great distances. In 1964, the discovery of a 10-foot-long predatory dinosaur called *Deinonychus* from the Kootenai near here helped change the perception of dinosaurs from plodding, cold-blooded reptiles to active, agile creatures more like non-flying birds. Now, most paleontologists think this dinosaur to be so close to birds that it was covered in feathers. By the start of the Late Cretaceous, about 100 million years ago, the Western Interior Seaway made a major incursion into the area, extending from the Arctic Ocean to the Gulf of Mexico. The sea deposited black marine shale of the Thermopolis Formation with fossils of ammonites and plesiosaurs.

Shells of oysters and cone-shaped belemnites (squid-like animals) that weathered out of the Jurassic Rierdon Formation are easily collected off the ground adjacent to the Crooked Creek Road.

White Cliffs of the Missouri River

Sandy Sediment of the Western Interior Seaway

Some of the best places to see rocks in Montana are from the water, particularly the White Cliffs section of the Missouri River Breaks, a magical landscape enjoyed by humans for millennia. The geology was first noted in the journals of Lewis and Clark in May 1805 and later surveyed by Ferdinand Hayden during his explorations of the Missouri River to Fort Benton in the mid-1850s. The area was designated a National Wild and Scenic River in 1976 and achieved national monument status in 2001. The float from Fort Benton to Judith Landing is 88.5 river miles, with the last 46 river miles passing through the White Cliffs section between Coal Banks Landing and Judith Landing.

Downstream from Fort Benton, the river cuts through Late Cretaceous shales and sandstones of the Marias River, Telegraph Creek, and Eagle Formations. They were deposited from 90 to 70 million years ago on a subtropical coastal plain east of the Sevier highlands and within the Western Interior Seaway, a north-to-south-trending ocean that connected the Arctic Ocean to the Gulf of Mexico. Sand was deposited by streams, on beaches, and as coastal sandbars, burying plants and dinosaur bones. Mud accumulated on the quiet ocean bottom farther offshore, which teemed with ammonites (squids) and clams. The dramatic White Cliffs, which rise up to 1,000 feet above the river, are composed of quartz sandstone of the Virgelle Member of the Eagle Sandstone. It was deposited over the ocean floor mud of Marias River Shale by coastal streams draining eastward as the ocean retreated. Cross-beds, the inclined internal layers in the sandstone, show that the streams flowed generally eastward toward the seaway.

The dramatic white cliffs of the Upper Missouri River are composed of resistant sandstone of the Cretaceous Virgelle Member of the Eagle Sandstone. —Courtesy of the Bureau of Land Management

As the Missouri River eroded down through the rising plains after the continental ice retreated, cliff walls formed as vertical cracks in the rocks widened each time water within them froze and expanded. Erosion of the underlying, softer shale undermined the cracked rock, which fell, leaving near-vertical cliffs and spectacular pillars and towers. Take time from the river to explore the slot canyons, formed by cracks widened by erosion. Look for unusual features in the sandstone, like pockmarks, pedestals, and concretions. Pockmarks form where surface water seeps into the porous sandstone and precipitates salt crystals that dislodge sand grains, making holes. Pedestal rocks, which look a bit like giant toadstools, form where resistant layers of iron-cemented sandstone protect the underlying, softer sandstone from erosion. Concretions are hardened, dark-colored balls of rock, some larger than cannonballs, that form when mineral cements precipitate outward from a nucleus, like a shell or bit of organic matter.

The White Cliffs section of the Missouri River float begins at Coal Banks Landing 12 miles south of Big Sandy. The 46-mile float takes four days and three nights. Depart the river at Judith Landing and return to Coal Banks Landing via MT 236 to Big Sandy.

Intruded into the white sandstone layers are several thin, vertical, dark-colored shonkinite dikes. The magmas were injected into vertical fractures in the sandstone around 50 million years ago. The igneous rock is much harder than the sandstone, so it weathers out in relief. These odd, dark-colored rocks are rich in iron-and magnesium-rich minerals like pyroxene but also contain minerals with alkali elements like sodium and potassium. The magma feeding these dikes likely came from volcanism to the northeast in the Bears Paw Mountains. One of the most famous pillars of shonkinite is Citadel Rock, described by William Clark on May 31, 1805, as a "high Steep black rock riseing [*sic*] from the waters edge."

Citadel Rock is an intrusion of shonkinite. —Courtesy of Mike MacLeod

Egg Mountain

Nests of the Good Mother Dinosaur

One of the world's most important dinosaur discoveries—nesting sites that confirmed nurturing behavior—is in Montana at Egg Mountain south of Choteau. As the story goes, in 1977, rock-shop owner Marion Brandvold discovered the first baby dinosaur bones found in North America. A year later, dinosaur paleontologist Jack Horner walked into her shop in Bynum, was taken to the site, and began the famous excavation of this dinosaur colonial nesting site, now called Egg Mountain. A sign on US 287 south of Choteau tells the story of Egg Mountain and provides a vista overlooking the Two Medicine Formation, the Late Cretaceous unit preserving the nests. The area beyond the fence is private land, so please stay on the road. It is illegal to collect vertebrate fossils on public land without a permit and is at the discretion and permission of the landowner on private land.

Layers of sandstone and shale of the Cretaceous Two Medicine Formation from the interpretative pullout on US 287 south of Choteau.

During Late Cretaceous time, rivers built coastal plain deltas into the Western Interior Seaway between 80 and 75 million years ago, depositing the Two Medicine Formation and the Judith River Formation farther east. Rising mountains from the Sevier tectonic collision loomed off to the west, with volcanoes producing ash that was carried downwind, spreading across the coastal plain and perhaps causing the demise of dinosaur herds. The fossil-rich middle member of the Two Medicine consists of stream and lake deposits exposed in a fold called the Willow Creek anticline. The nesting sites occur in the mudstones, which were deposited on floodplains adjacent to sandy rivers and lakes. The

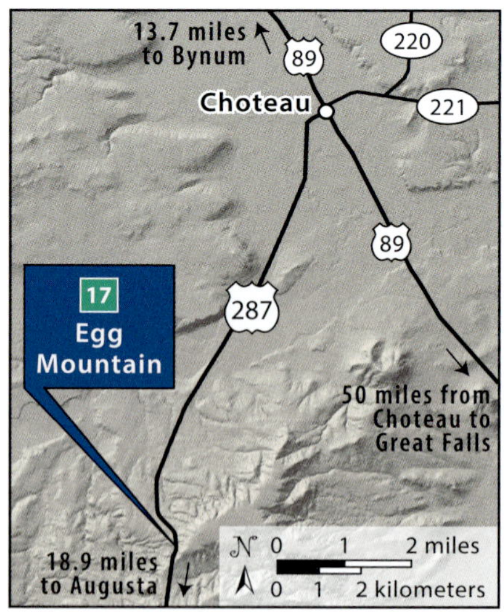

About 7 miles south of Choteau, look for a pullout on US 287.

coastal plain was also home to a variety of other animals, including pterosaurs, turtles, mammals, insects, and freshwater clams and snails.

Egg Mountain was where the duck-billed dinosaur *Maiasaura peeblesorum*, meaning "good mother lizard," nested and cared for its young. Prior to this discovery, scientists had not yet demonstrated that some dinosaurs nurtured their young. Since the initial discovery, hundreds of specimens in all stages of life, from dinosaur eggs and embryos to full-grown adults, have been found, providing an unparalleled understanding of the behavior of this dinosaur. They lived in herds of thousands on the coastal plain, migrated seasonally for food, nested as a group, and raised their young until the entire herd could again move on. Studies of coprolites or fossil feces show that these plant eaters both grazed and browsed. Many other types of dinosaurs are known from the Two Medicine Formation, including a small, meat-eating dinosaur named *Troodon*, which also nested at Egg Mountain. Its eggs indicate brooding behaviors very similar to modern birds.

There are many excellent paleontological museums in Montana, a list of which can be found on the Montana Dinosaur Trail website (https://mtdinotrail.org). The Montana Dinosaur Center in Bynum, north of Egg Mountain and worth a visit, has great displays and serves as a dinosaur research center. The Museum of the Rockies in Bozeman hosts many of the fossils from Egg Mountain.

Nest of Troodon, a small meat-eating dinosaur from the Two Medicine Formation. Notice the nest rim with its fossil egg clutch under the white plaster. —Courtesy of Dave Varricchio

This 4-inch-long theropod egg fossil from Egg Mountain is on display at the Montana Dinosaur Center in Bynum.

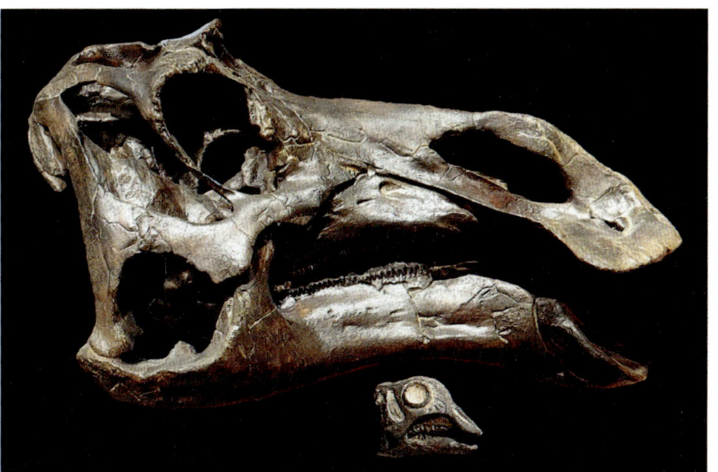

The skulls of an adult and hatchling Maiasaura peeblesorum (Montana's State Fossil) from the Late Cretaceous Two Medicine Formation. The adult skull is 28.5 inches long and the hatchling only 3 inches long. —Courtesy of Scott Williams, Museum of the Rockies

18 Sun River Canyon

Thrust Faults of the Rocky Mountain Front

The Rocky Mountain front rises dramatically from the plains in Montana, evoking the immensity of time and space. The mountains are great slabs of Mesoproterozoic and Paleozoic rocks that were shoved eastward, up and over younger, softer, more easily eroded, Cretaceous sedimentary rocks of the plains. Sun River Canyon, west of Augusta, is a great place to see the giant slabs of rock that started moving eastward at the very end of Jurassic time, as North America began colliding with tectonic plates to the west. The deformation, part of the Sevier mountain building, ends just a bit east of where the mountains rise from the plains.

The mountains seem to grow with each mile as you drive west from Augusta on Sun Canyon Road. The Sawtooth Range ahead seems impenetrable until you see the narrow gorge eroded through the great wall of limestone, an opening into the Sun River Canyon. The west-dipping slab of limestone is the fossil-rich, tropical marine limestone of the Mississippian Madison Group. It was not deposited here, but rather, it was pushed on a thrust fault that is visible across the river where the impressive slab of limestone rests on dark-colored Cretaceous sandstone and shale. This slab is the easternmost overthrust slab in the Sevier fold-and-thrust belt. It's also the youngest of the many big thrust faults because the faults propagated from west to east. Smaller

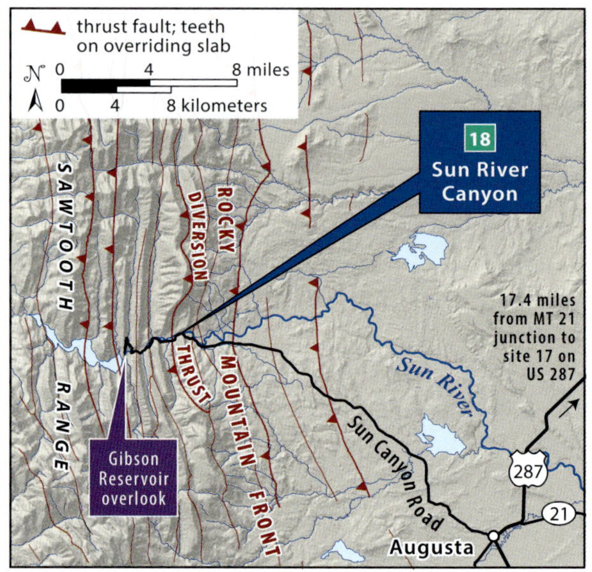

From Augusta, travel 66 miles west on Sun Canyon Road to the entrance of Sun River Canyon, and continue with stops to Gibson Reservoir Overlook.

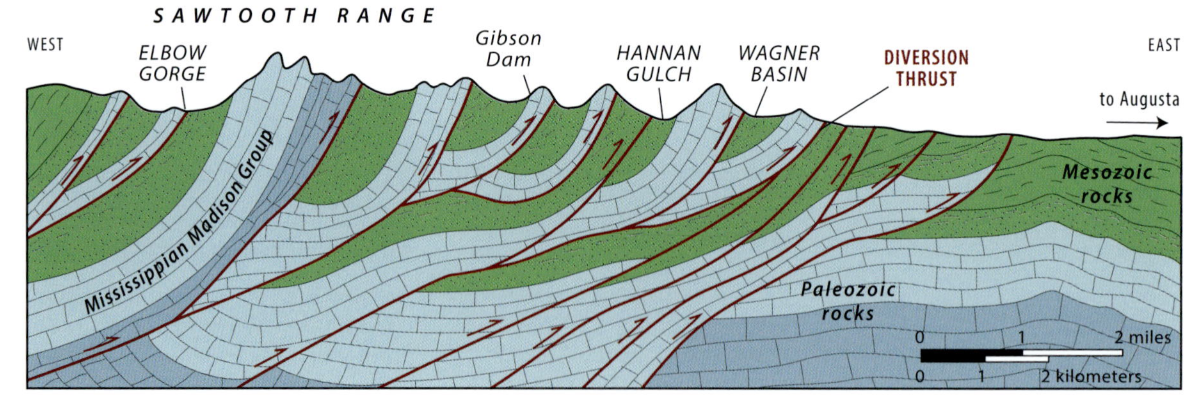

West to east cross section along the line of the road to Gibson Dam illustrating the many slabs of Madison Group limestone that were thrust up and over younger Jurassic and Cretaceous sedimentary rocks. —Modified from Hyndman and Thomas, 2020

A close-up of the thrust fault plane shows that it consists of broken up rock called fault gouge, which formed as movement along the fault crushed and ground up the rock.

folds and thrusts continue eastward into the plains, forming a disturbed belt of easily eroded Jurassic and Cretaceous sedimentary rocks.

As you drive through the canyon, the landscape widens into Wagner Basin, an area where less-resistant Jurassic and Cretaceous sandstone and shale were deposited on an eroded surface of the Madison Group limestone. The valley narrows again, passing through a second slab of limestone that slid up and over the Cretaceous shale on a slightly older thrust fault. This second slab is unconformably overlain by more Jurassic and Cretaceous sandstone and shale forming Hannan Gulch. Continuing west, the road passes through a third narrow gorge eroded through another thrust slab of Madison Group limestone, this fault older than the other two and forming the anchor for Gibson Dam. Dating of the clays that formed from the heat generated during thrust faulting indicates they were active 67 million years ago. A close look at the fault planes shows that the rocks are totally broken up into fault gouge, formed by the grinding and crushing of rock during movement along the faults.

The Diversion thrust fault at the entrance to Sun River Canyon places Mississippian Madison Group limestone over Jurassic and Cretaceous sandstone and shale.

The folds and faults exposed on the Rocky Mountain front must have formed at depth in Earth's crust because they require much higher pressure and temperature than exists at the surface. We see them today because the front and plains are rising together and eroding. The hard limestones of the mountains resist erosion better than the soft shales underlying the plains, so they stand out in relief.

19 Chief Mountain

Carried East on the Lewis Thrust

Rising more than 4,000 feet above the plains, Chief Mountain demands attention and has long played a prominent role in the cultures of indigenous people. The Blackfeet, who call it Ninaistaki, consider it the sacred home to the creator of thunder. The peak appeared on early maps of the area, and in 1901, Bailey Willis of the US Geological Survey recognized that Chief Mountain had moved east many tens of miles along the Lewis thrust. This low-angle compressional fault ramped up from great depth, carrying 1.47-billion-year-old Belt Supergroup rocks eastward over Cretaceous shale during the Sevier mountain building. Thrust faults place older rocks over younger rocks, but the age difference between the older and younger rocks here—over 1.4 billion years—is amazing. The overlying rock is sandy dolomitic limestone of the Belt's Altyn Formation (and overlying Appekunny Formation), and the underlying rock is shale and sandstone of the Late Cretaceous Two Medicine Formation.

The Lewis thrust moved horizontally on weak Cretaceous shales, slowly transporting a slab of rock that was at least 300 miles long and 2 miles thick. It traveled about 80 miles from west to east, stopping about where the mountains now rise from the plains on the east side of Glacier National Park. Based on dates from cross-cutting volcanic rocks, the fault moved between 75 and 59 million years ago. All rocks above the fault are gently folded into a huge downfold called the Akamina syncline.

The Lewis thrust plane is nearly horizontal, so friction on its surface must have been very low to facilitate the movement. Perhaps expandable clays in the underlying Cretaceous shales dropped friction dramatically. The fault moved shortly after deposition of the shales, so they might have been wet, and when loaded by the heavy Belt Supergroup rocks, pore water might have squeezed out of them and lubricated the fault plane. The Cretaceous rocks below the fault range from seemingly undeformed rocks that parallel the fault to intensely deformed rocks. Over time, the fold-and-thrust belt continued to migrate east of the Lewis thrust but mostly deforming Cretaceous rocks near the top of the sedimentary pile. The thrusting ended around 56 million years ago.

Erosion of the overlying Belt rocks above much of the fault plane isolated Chief Mountain, forming a klippe, a German word meaning "cliff." The great thrust slab might

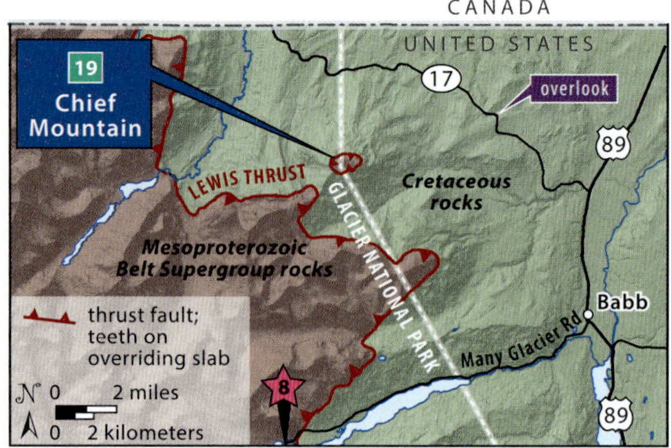

From Babb, travel 4 miles north on US 89, and turn left (northwest) on MT 17 (Chief Mountain Highway). Travel 5 miles to the Chief Mountain Overlook.

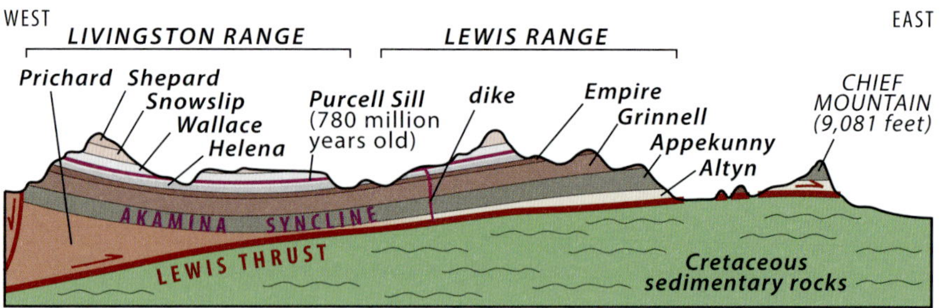

Simplified west to east cross section through Glacier National Park showing the Lewis thrust and the erosional isolation of the Chief Mountain klippe. —Hyndman and Thomas, 2020

Chief Mountain (Ninaistaki) viewed from the Chief Mountain Highway north of Babb. Notice the landslide scar on the northeast face caused by an earthquake in 1992.

have extended a bit farther east, but any evidence has been eroded away, and Chief Mountain is most likely very close to the eastern front of the Lewis thrust. Exposure of this once deeply buried fault shows the ground was raised after it stopped moving, likely due to Cenozoic crustal stretching and thermal uplift in western Montana. Miles of the overlying rock were eroded away. The large scar on the northeast face of Chief Mountain formed when weakened rock collapsed during an earthquake on July 2, 1992, showing that the landscape is still a work in progress.

Bannack State Park

Deposition of the Beaverhead Conglomerate

As we have learned, the Sevier mountain building is simply a term used to describe a time and style of crustal shortening that created the fold-and-thrust belt in western Montana. The westward movement of North America during the breakup of Pangea created a convergent plate boundary along the continent's western margin, and the compression shoved thick slabs of sedimentary rocks to the east. The crustal shortening was primarily active from Cretaceous through earliest Paleogene time. The thrust front is particularly complex in southwest Montana, where sediments that were shed off the rising mountains were deformed during or shortly after deposition.

From MT 278, travel 3 miles south on the Bannack Bench Road. Stay to the left onto Bannack Road for 0.7 miles to the parking lot. Check in at the visitor center, and walk to the east end of town. Follow the dirt road east along Grasshopper Creek about 2 miles to where it cuts a deep canyon into the red-colored Beaverhead conglomerate. —Faults from Mosolf and others, 2025

Southeast of the gold-mining ghost town of Bannack State Park, a dirt road follows Grasshopper Creek downstream into a canyon cut into the Late Cretaceous Beaverhead Group conglomerates. At the canyon entrance, an east-dipping thrust fault places Madison Group limestone (on the northeastern skyline) over the Beaverhead conglomerate. The conglomerates are loaded with mostly angular pebbles and cobbles of Madison limestone that were not transported very far from their sources. The conglomerates are interbedded with brick-red sandstone with indistinct layering, and some fill deep, steep-walled channels, features typical of deposition on alluvial fans, especially those adjacent to steep, tectonically active mountain fronts.

The thrust fault, called the Ermont thrust, shoved the Mississippian Madison Group limestone over the conglomerate while it was being deposited. We know this because the fans were tilted and steepened during deposition, then eroded before more fan sediments buried them as the fans migrated away from the rising mountains and into the foreland basin. Careful field work shows it happened at least three

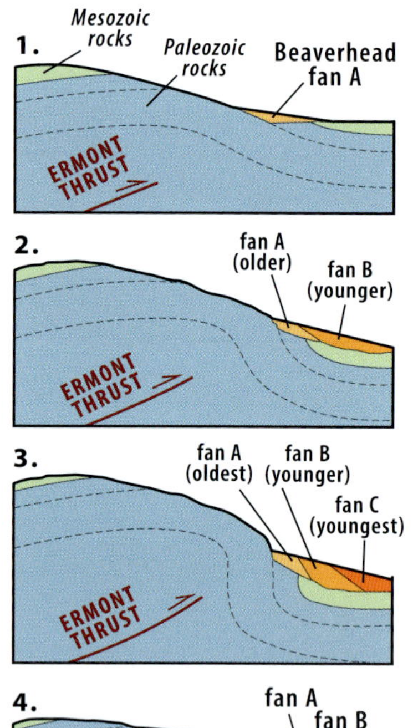

Interpretation of how alluvial fan deposits were deformed during active tectonism. —Modified from Azevedo, 1993

times during sudden fault movements that instantly tilted the fans. Perhaps the steep-walled channels formed when the fans suddenly steepened and were quickly eroded. Breccias are present that may have been deposited when earthquake shaking caused rocks to fall on the fans. Once the thrust slabs migrated over the fans, the cobbles served as ball bearings, facilitating movement. The weight of the overriding slab fractured and squashed the cobbles in the process.

Not long after the thrust was emplaced, granitic magma intruded the limestone just above the thrust and turned it into marble and skarn, a rock rich in ore minerals. Placer gold—concentrations of the heavy mineral in stream deposits—was discovered here in 1862, followed by hard-rock mining of gold-bearing quartz deposits near the intrusions. A few old mine dumps are visible below the Bannack granodiorite southwest of Grasshopper Creek.

Beaverhead Group alluvial fan deposits in Grasshopper Creek canyon include steep-walled channels (gray) cut into red sandstone. The channels are filled with rock debris of Madison limestone and likely formed when alluvial fans were tectonically steepened. The inset shows the many pebbles and cobbles of gray Madison limestone in the conglomerate.

View of the Ermont thrust looking north from Grasshopper Creek downstream (east) of Bannack. In Cretaceous time, Mississippian Madison Group limestone was shoved over alluvial fan deposits of the Cretaceous Beaverhead Group as they were being shed from the rising mountains.

21 Blodgett Canyon

Granite of Little Yosemite

As sedimentary rocks were being shuffled eastward and stacked by Sevier-style compression in Cretaceous time, the mountains grew tall, and Earth's crust thickened. Granitic magmas form in extremely thick crust because of the high temperatures and pressure. In central Idaho, a large body of magma crystallized at depth into a mass of mostly granitic rocks known as a batholith. The Idaho batholith is divided into two main bodies, the older Atlanta lobe (83 to 67 million years) to the south and the younger Bitterroot lobe (66 to 53 million years) to the north, some of which is exposed in the Bitterroot Mountains of western Montana. The most picturesque place to see these intrusive igneous rocks is in Blodgett Canyon west of Hamilton, known to some as Little Yosemite.

The Bitterroot lobe crystallized into granite and granodiorite about 10 miles below the surface while in contact with Belt Supergroup and older rocks. Magmas can erupt at the surface creating volcanoes, but there is no evidence that the Bitterroot granitic magma ever erupted. The Bitterroot lobe is composed of many different intrusions that vary in age and composition, with an overall sheet-like shape only a few miles thick.

In Blodgett Canyon, the rocks along the trail are mostly mica-rich granite and granodiorite, composed of glassy quartz, white feldspars, light-colored muscovite mica, dark-colored biotite mica, and some long, thin crystals of black hornblende. We can see the minerals because the magma cooled slowly, far enough below the surface to retain heat and allow the crystals time to grow large enough to see them with the unaided eye. Some of the white feldspar crystals are

N 0 2 4 miles
0 2 4 kilometers
49 miles from Hamilton to Missoula and I-90
KOOTENAI CANYON
34
Stevensville
BITTERROOT RANGE
IDAHO
MONTANA
21 Blodgett Canyon (trailhead)
Bitterroot River
93
BLODGETT CANYON
Hamilton

From US 93 in Hamilton, travel 2.8 miles west on West Main Street to Westbridge Road and then to Canyon Creek Road. Turn left (west) on Blodgett Camp Road, and travel 2.8 miles to the Blodgett Trailhead. Follow the trail for about 3 miles west for the best views.

The glacially sculpted peaks are now exfoliating or peeling like an onion as thin slabs of rock expand, crack, and fall, creating piles of talus or fallen rock below.

larger, about the size of a dime, and have a distinct crystal shape, showing they grew early while still surrounded by magma. The smaller minerals formed later, competing for space, and so do not have crystal shapes. The rocks near the mountain front show distinct mineral layering as a result of crustal stretching and the formation of intensely sheared rocks, called mylonite, in Eocene time.

Since their formation at depth, these rocks experienced a great deal of uplift and removal of overlying rock. Blodgett Canyon boasts a U-shape profile and horn-shaped peaks sculpted by glacial ice in Pleistocene time. Since the ice retreated around 12,000 years ago, the granitic rock has been exfoliating or peeling like an onion as it expands, cracks, and delaminates into thin, unstable slabs of rock. At least two mechanisms are at work. First, as the overlying rocks were eroded away, the pressure was released on the formerly deeply buried granitic rocks. Second, the planar fabric of the mylonite provides an inherent weakness to the rock. Freezing and thawing of water in the cracks, called frost wedging, aids in the process, causing rocks to fall and accumulate as talus below the peaks.

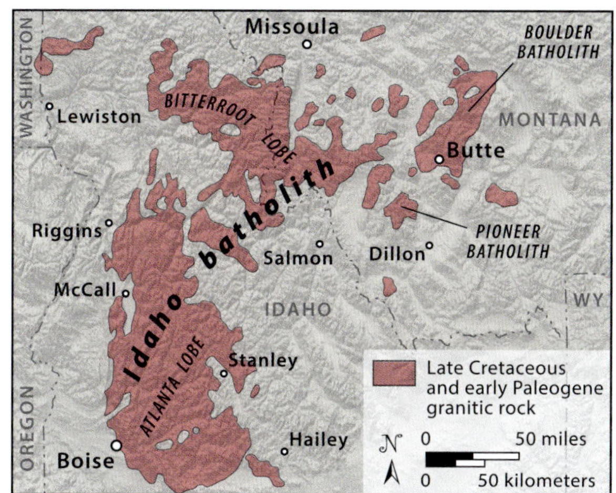

Distribution of the intrusive rocks of the Idaho batholith and similar granitic intrusions to the east in Montana.

The Blodgett Canyon Trail west of Hamilton provides spectacular views of the Bitterroot lobe of the Idaho batholith, granitic rocks intruded into sedimentary rocks of the Belt Supergroup between 66 and 53 million years ago.

Typical granitic rock of the Bitterroot lobe of the Idaho batholith exposed along the trail in Blodgett Canyon.

22 Humbug Spires

Weathered Cracks in the Boulder Batholith

During Cretaceous and earliest Paleogene time, plate collision and crustal thickening to the west produced the large volume of granitic rocks now exposed in central Idaho and the Bitterroot Mountains. It also produced granitic magma that intruded the Sevier fold-and-thrust belt, injecting into folds and faults and erupting onto the surface. The largest body of granitic rock is the Late Cretaceous Boulder batholith, formed between 80 and 70 million years ago but mostly crystallized around 76 million years ago. The large batholith, exposed over 1,700 square miles between Butte and Helena, is composed of fifteen distinct intrusions, called plutons, some of which are sheetlike and emplaced only a few miles below the surface. The magma that reached the surface erupted the enormous Elkhorn Mountains volcanic pile. Some eruptions were explosive, producing fast-moving currents of hot gas and volcanic matter that killed dinosaurs and knocked down trees. In Eocene time, crustal stretching moved the batholith more than 12 miles to the east of its original location on a deep, low-angle fault.

From Melrose, travel 5.9 miles north on I-15 to Moose Creek Road (exit 99). Take a right (northeast), and follow the improved dirt road for 3.4 miles to the Moose Creek Trailhead. Granitic rock, boulders, and grus abound along the trail, with the first prominent spire visible at 3 miles. —Map of granitic bodies from McDonald and others, 2012

There are many special places to see this batholith but none more so than the Humbug Spires Wilderness Study Area south of Butte near Divide. This popular rock-climbing area boasts more than fifty rock spires, some reaching 600 feet above the surrounding terrain. They are composed of granodiorite of the Moose Creek pluton, which intruded a few miles below the surface around 74 million years ago. The granite-like rock consists of crystals of white feldspar, gray quartz, black biotite mica, and stubby, black needles of hornblende. Within the granodiorite are inclusions of rounded blobs of dark igneous rocks that were either picked up by the magma or crystallized from blobs of unmixed magma. Multiple episodes of crustal stretching and erosion over the last 50 million years brought the rocks to the surface, where weathering of fractures in the rocks formed the rock spires.

The Boulder batholith is so named because intersecting fractures weather into spheroidal blocks that look like boulders. Most of the batholith has two or three fracture directions that intersect at 90 degrees to one another. They formed for multiple reasons. Perhaps the rock contracted as it crystallized because magma shrinks about 10 percent as it cools. The solid granite might have also cracked when flexed upward from folding in Cretaceous time or stretching in Cenozoic time. The more horizontal fractures likely formed

Inclusions of dark igneous rock were either picked up by the magma or crystallized from blobs of unmixed magma.

Fins exposed along the trail show how nearly vertical fractures in the granitic rock weather into spires over time.

as the rock was raised to the surface, expanding and cracking with pressure release. Some of the nearly vertical fractures are filled with dikes of granitic rock, suggesting these cracks formed shortly after the pluton crystallized and the remaining hot fluids spread into them. In the Moose Creek pluton, these vertical fractures are more widely spaced, with few cross fractures, leaving the hard rock to weather into spires rather than boulders, like elsewhere in the batholith.

The spires formed when rainwater, which is slightly acidic, seeped into the steeper fractures, weathering the minerals and widening and deepening the cracks. Minerals like feldspar weather to clay, while platy micas expand and contract with wetting and drying, breaking rock into granules, called grus, that are easily eroded. Water freezing in the cracks widens them as well, and with time, the vertical slabs of resistant rock become widely separated and stand out in relief from the surrounding weathered rock and soil, forming a rock spire.

A beautiful spire composed of granitic rock of the Cretaceous Moose Creek pluton serves as a reward for about 3 miles of easy hiking on the Moose Creek Trail.

23 Farlin Ghost Town

Treasure in the Pioneer Mountains

Montana is called the Treasure State because of its mineral wealth. Some of that treasure formed during Cretaceous time, when granitic magmas intruded sedimentary rocks and formed numerous ore-bearing deposits. As prospectors headed west, they initially panned the creeks in search of gold, which they first found at Gold Creek near Drummond in the spring of 1858. A few years later, discoveries at Argenta, Bannack, Alder Gulch, Butte, and many other locations drove a rapid migration into Montana. Once the sources of the panned minerals were discovered, hard-rock mining began. In the Pioneer Mountains west of Dillon, a claim was staked near Birch Creek in 1864. By the late 1800s, the mining camp of Farlin was a booming town of more than five hundred people with a post office, schoolhouse, and smelter, some of which still stand today along the Birch Creek Road.

During Cretaceous time, tectonic plate collisions to the west formed granitic magmas, like the Pioneer batholith near Dillon, that were injected into the fold-and-thrust belt. This 300-square-mile, sheet-like mass of mostly granodiorite (a granitic rock) consists of fifteen smaller intrusions, the largest of which was intruded around 72 million years ago into older sedimentary rocks just after they were deformed. At Farlin, the granodiorite intruded the Mississippian Madison Group limestone, which was heated and chemically altered as hot fluids exchanged dissolved elements. The process formed skarn, a metamorphic rock where carbonate minerals were replaced by calcium-rich silicate minerals. The process also enriched the rock in ore-bearing minerals that were mined primarily for copper, silver, lead, iron ore, and gold.

Crustal stretching and uplift beginning in Eocene time, along with erosion, have exposed a superb cross-sectional view of the metamorphic zone between the granodiorite and the limestone. To see the zone, start at the picnic area west of Farlin, and walk east. The granodiorite magma was injected into the limestone several miles below the surface, where it cooled. It looks layered due to pressure-release fractures formed as it was raised to the surface. The rock is composed of visible crystals of quartz, feldspar, biotite mica, and hornblende. As you near the bake zone, the granodiorite is altered and contains abundant ore-bearing minerals like brown-colored grossular garnet, green malachite, and blue azurite.

From Dillon, travel 12.1 miles north to Apex (exit 74), and turn left (west) on the Birch Creek Road. Travel 7 miles, and park at the picnic area. Walk east on the road to see the contact metamorphic zone on the left (north) and slag pile on the right (south).

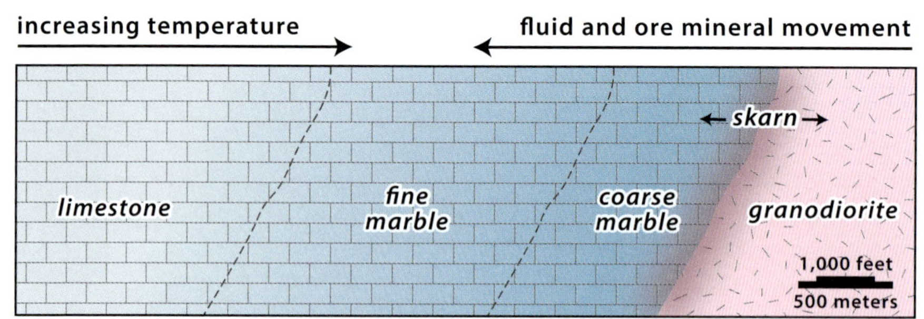

Simplified sketch of a contact metamorphic zone with skarn mineralization caused by the exchange of elements by hot fluids. —Modified from Virginia Sisson, 2025

The skarn is orange to brown in color due to the oxidation of iron-bearing minerals. It's easy to see on the slope because of the waste piles from past underground mining. The primary ore-bearing minerals are copper sulfides like chalcopyrite, chalcocite, and bornite; copper carbonates like malachite and azurite; lead-rich galena; iron-rich magnetite, and deposits of native copper, silver, and gold. The skarn transitions into white marble, a limestone that was baked but not mineralized. A bit farther away from the source of heat is unmetamorphosed limestone with marine fossils of crinoids and corals.

To extract the ore from the skarn, the rock was heated, or smelted, on site, leaving behind a pile of glassy smelter waste, called slag, next to Birch Creek. The pile of glassy, black slag has ropy flow structures and large, rounded slag balls that were dumped from kettles into the slag pile. Water-quality testing shows that the slag has not contaminated Birch Creek, probably because glass is resistant to chemical breakdown in the natural environment.

The Birch Creek skarn is rusty orange from the oxidation of iron-bearing minerals like hematite and magnetite. The green mineral is malachite, a copper-bearing mineral.

Molten waste from a smelter was poured into the creek where it quickly cooled, forming glassy slag. Look for flow features and a metallic sheen from waste minerals like pyrite.

The orange-colored waste piles show the trend of the Birch Creek skarn up the north side of the Birch Creek drainage. The granodiorite of the Pioneer batholith is exposed to the left (west) and the unaltered Mississippian Madison Group limestone to the right (east).

24 Crystal Park

Digging for Vein Quartz in Weathered Granitic Rocks

There are few places to take your family on a better geological excursion than Crystal Park in the Pioneer Mountains south of Wise River. The drive along the Pioneer Scenic Byway is an alpine wonderland of glaciated peaks, broad glacial valleys with meandering streams, and alpine meadows or parks that were once covered by a huge ice cap during the Pinedale glaciation. The park is ADA accessible, so people of all ages and abilities can easily collect quartz crystals; plenty can just be picked up off the ground.

The quartz crystallized in cracks in granitic rocks of the huge Pioneer batholith, which is composed of many smaller intrusions. The quartz crystals occur in the Uphill Creek intrusion, a granodiorite that intruded deformed sedimentary rocks of the Sevier fold-and-thrust belt during Cretaceous time, around 72 million years ago. The granodiorite cooled slowly at depth, allowing enough time for crystals to grow big enough to easily see, including glassy quartz, white feldspar, black hornblende, and dark, platy biotite mica. This deeply formed rock is exposed today due to 50 million years of crustal stretching, uplift, and erosion of the Pioneer Mountains.

Soon after the magma crystallized, hydrothermal fluid, which is superheated water containing silica, permeated cracks and other open spaces in the granodiorite and precipitated crystals of vein quartz as the fluid cooled. The open spaces allowed the quartz crystals to grow into their ideal shape of six-sided prisms terminated with six-sided pyramids, typically at one end. The crystals range in size from smaller than your little finger to several inches in diameter. They can be clear, white, gray (smoky), or purple (amethyst); many have tiny inclusions of other minerals or small bubbles of fluid. Some are stained red or orange from iron oxide in the groundwater. Some crystals occur as scepter quartz in which a second generation of crystal tips (usually larger) grow on

From the turnoff to Elkhorn Hot Springs, travel 5.3 miles north on the Pioneer Mountain Scenic Byway to Crystal Park. Walk the paved switchback trail to access collecting sites.

A 0.8-inch-long, pale amethyst quartz crystal scepter collected from Crystal Park. —Courtesy of Richard Gibson

top of another quartz crystal. Crystals that have the shadowy outline of a crystal within a crystal (phantoms) are rare prizes.

Once exposed, the granodiorite, with its quartz-filled veins and cavities, was intensely weathered into a granitic sand called grus. With repeated wetting and drying, the platy mica grains expanded and contracted, busting the rock into granules that are easy to dig with a shovel. Chemical weathering turned the feldspar crystals into clay, which also helped to break down the granodiorite. If you look closely, you might see circular remnants of less altered granodiorite surrounded by intensely weathered rock, showing that weathering along intersecting fractures played a big role in the breakdown of the rock. Surface erosion each year washes away the grus and leaves the more resistant crystals of quartz lying on the surface, just waiting to be picked up, no tools required. Many visitors come loaded with picks and shovels ready to dig deep, but don't be intimidated if all you have is a plastic sandwich bag and a keen eye; you will also be rewarded.

Geared up and ready to go!

The granitic rock, heavily weathered to grus, is easy to dig.

Collecting quartz crystals at Crystal Park is great fun for kids of all ages. Such experiences help instill a passion for learning about the natural world that is desperately needed, now more than ever.

Tower Rock State Park

Explosive Volcanoes of the Adel Mountains

Students commonly ask if the Cretaceous granitic rocks in Montana were once magma chambers beneath erupting volcanoes. Except for the Elkhorn Mountains volcanic field, which erupted from the Boulder batholith between Butte and Helena, there isn't evidence for many volcanoes of this age here. Perhaps the magma didn't erupt, or perhaps time, uplift, and erosion erased the evidence. The eroded remains of at least one Cretaceous volcano, however, is well exposed in a canyon cut by the Missouri River in the Adel Mountains between Craig and Tower Rock State Park southwest of Cascade.

The Adel Mountains volcanic field erupted between 76 and 73 million years ago, leaving volcanic deposits over an area of about 350 square miles. Most of the rock was extruded as lava, deposited as volcanic sediment, or crystallized in the volcanic vents. There are thousands of dark-colored, coarse-grained dikes that radiate out from the vents, some extending more than 22 miles from their source and terminating in dome-shaped laccoliths, blisters of intrusive rocks formed as the magma injected parallel to the sedimentary layers. Crown, Shaw, Square, and Cascade Buttes west of Great Falls are laccoliths formed at the end of these dikes, now exposed by erosion of the surrounding, softer sedimentary rocks.

Tilted layers preserve the remnant slopes of deeply eroded volcanoes built of layers of lava, ash, and rock debris, like the big stratovolcanoes that form above the subduction zone in the Pacific Northwest today. The lava flows are basaltic andesite, a dark, fine-grained rock rich in the minerals plagioclase

From Craig, travel 15.5 miles northeast on Old US Highway 91 through the canyon of the Missouri River to Tower Rock State Park. The Adel Mountains volcanic vents produced thousands of dikes that radiate outwards up to 22 miles and terminate in laccoliths like Shaw Butte near Simms. —Dikes from Harlan and others, 2005

A shonkinite intrusion exposed along the frontage road is a rare igneous rock that combines potassium feldspar and pyroxene, the large black crystals.

Angular volcanic fragments of andesite and shonkinite float in a muddy matrix, showing this viscous debris flow, or lahar, did not travel very far from its source volcano.

View downstream from the frontage road at the Missouri River canyon cut through the Cretaceous Adel Mountains volcanic rocks near Craig.

feldspar and pyroxene. At the surface, lava cools quickly, forming rocks with non-visible crystals. Some rocks, however, are loaded with visible crystals that grew in the magma chamber before erupting onto the surface or when intruding at shallow depths. The lava is mostly interbedded with ash, volcanic breccia, and sediments. The breccias, rocks composed of large angular rock fragments in a fine-grained groundmass, formed when explosive eruptions caused viscous flows of water and rock debris called lahars. The volcanic rocks piled up on Cretaceous sediments, and as the pile grew, its weight bowed them down, forming radial paths for the magma to squirt out of vents and crystallize into dikes and laccoliths of shonkinite, a dark-colored igneous rock rich in the minerals potassium feldspar and pyroxene.

The origin of the Adel Mountains volcanic field is unclear. It was far from any subduction zone that might have existed in the Late Cretaceous. Because some of the volcanic field overlapped the active eastern edge of the Sevier fold-and-thrust belt, perhaps the tectonic disruption provided a way for the magma to reach the surface. The crustal shortening likely pushed the upper mantle deeper into Earth, where it melted at higher pressure and made shonkinite.

26 Beartooth Front near Red Lodge

Laramide Uplifts and Limestone Palisades

During late Cretaceous time, compressional faulting transitioned from thin-skinned Sevier deformation to thick-skinned Laramide deformation, with steep faults emerging from deep within the basement and raising large blocks of it to the surface. There are many good examples of Laramide uplifts in Montana, but the one with the most geological traffic is the Beartooth uplift near Red Lodge. The Yellowstone Bighorn Research Association (YBRA) geology field station, along the mountain front south of town, was founded by Princeton University in 1936. Each year, geology students from across the country map the Beartooth front, including the complex Beartooth fault and the steep layers of sedimentary rock draped over this uplift. The Beartooth fault dips about 35 degrees to the southwest and, in Late Cretaceous to early Paleogene time, moved the block of basement to the northeast. The amount of movement on the fault was huge, and it is estimated to have caused as much as 20,000 feet of offset. It is well exposed to the northwest of the YBRA field station and along the Meeteetsee Trail Road south of Red Lodge.

The core of the massive Beartooth uplift consists of hard, Archean-age granite and metamorphic rocks that are among the oldest rocks on Earth (site 1). During the Laramide deformation, the Paleozoic and Mesozoic rocks were drape-folded over the rising block of older rock. The younger sedimentary rocks were eroded from the top of the block, except for a remnant atop the block at Beartooth Butte in Wyoming, but steeply dipping beds of the folded rock remain along the Beartooth front. The most visually striking are the resistant limestone palisades or fins.

From Red Lodge, travel 4.5 miles south on US 212 to Howell Gulch Road. Turn left (east), and travel 1.2 miles uphill to the Yellowstone Bighorn Research Association geology field station to see the tilted sedimentary rocks of the Beartooth front. Limestone palisades are also well exposed at Red Lodge Mountain Ski Resort and at North Fork Grove Creek about 8.6 miles south of Red Lodge on the Meeteetsee Trail Road.

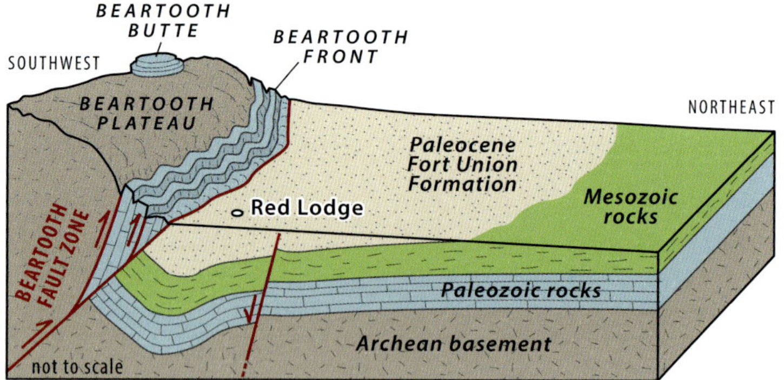

Simplified block diagram of the Beartooth uplift as it looks today. From Late Cretaceous to early Paleogene time, the Beartooth fault raised the basement and draped the Paleozoic rocks over the block. Erosion has since disconnected Paleozoic rocks atop the block at Beartooth Butte from those dipping steeply along the Beartooth front. —Modified from Hyndman and Thomas, 2020

View of the Beartooth front looking northwest from the Howell Gulch Road near YBRA. Archean basement rocks were thrust up during Laramide deformation. The overlying Paleozoic sedimentary rocks were drape-folded over the rising basement block, forming the steeply dipping, limestone palisades.

Particles of the eroded Paleozoic and Mesozoic rocks were deposited in the adjacent foreland basin as alluvial fan sediments of the Fort Union Formation, a Paleocene unit from the early Cenozoic Era. The fan sediments were folded and overrun by the Beartooth fault. The Fort Union records the unroofing history of the rising block, with the youngest rocks eroded off the uplift deposited at the bottom of the debris pile, like deconstructing a building from the top down. We know that fault movement caused big earthquakes because the foreland basin deposits are loaded with disrupted bedding caused by shaking in wet sediments.

Steeply dipping limestone palisades of Ordovician Bighorn Dolomite in the North Fork Grove Creek drainage on the northeastern front of the Laramide Beartooth uplift south of Red Lodge.

Makoshika State Park

Extinction of the Dinosaurs

As the Mesozoic Era came to an end, so did the reign of the nonavian dinosaurs. Since Jurassic time, dinosaurs had thrived along the streams, lakes, and coastal environments of the foreland basin to the east of the mountains. Their existence over a hundred million years is arguably much more important than their death, but we love a good murder mystery, and the mass extinction that took away 75 percent of all species on Earth is one for the ages. Known as the Cretaceous-Paleogene (K-Pg) boundary, this global event profoundly changed our planet 66 million years ago. The boundary is exposed across eastern Montana, and Makoshika State Park near Glendive provides a truly special place to see it.

A mass extinction is defined as being geographically widespread, involving a variety of marine and nonmarine species, and taking place over a short period of time, say less than a few million years. The Cretaceous-Paleogene boundary checks every box and contains very important "who done it" clues. At Makoshika State Park, the Cretaceous-Paleogene boundary is marked by a clay layer within a coal bed between the Cretaceous Hell Creek Formation and the Paleogene Fort Union Formation. Both were deposited in subtropical climates by streams and in swamps near the Western Interior Seaway, spanning the Cretaceous-Paleogene boundary from 67 to 64 million years ago. Below the coal bed are the remains of *Triceratops* and *Tyrannosaurus rex*, but above it, they are gone.

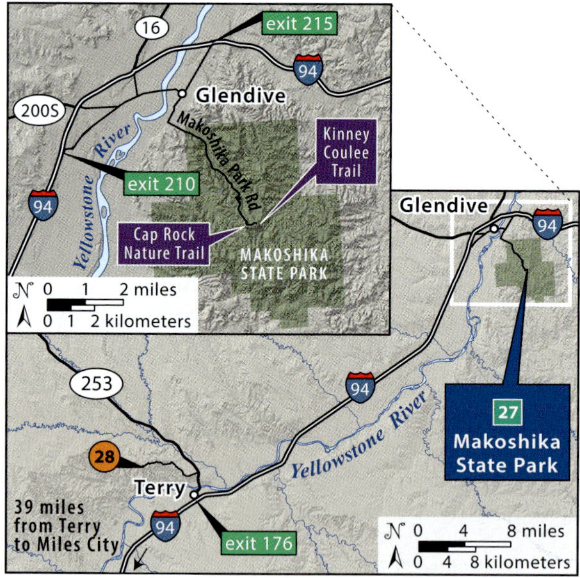

From South Taylor Avenue in Glendive, travel a half mile southeast on Snyder Street to Makoshika State Park Visitor Center. Travel Makoshika Park Road to access walking trails. Consider the Cap Rock Nature Trail and the Kinney Coulee Trail to see badlands topography and the Cretaceous-Paleogene extinction boundary.

A cap of iron-cemented, stream-deposited sandstone protects the underlying, softer sandstone to create this pillar.

The evidence suggests that 66.04 million years ago, a large meteorite crashed into what is now the Yucatán Peninsula of Mexico, engulfing Earth in a shroud of scorching debris that set the planet's vegetation on fire, caused nitric acid rain, and blackened the skies. The cloud of debris and ash disrupted photosynthesis and cooled the planet for decades. The clay layer in the coal bed contains the incriminating evidence for an extraterrestrial culprit, including soot from global wildfires; high concentrations of iridium, a rare element on Earth but highly enriched in the orbiting asteroids and meteorites of our solar system; and grains of shocked quartz, which have distinctive fractures formed by hypervelocity impacts at speeds over Mach 8 (more than 6,000 miles per hour). In the layers immediately above the clay fallout layer, a spike in fern spores shows this pioneering plant spread rapidly after the calamity. There are other proposed causes for the extinctions, but regardless, some animals survived, and the nonavian dinosaurs perished.

The extinction boundary is exposed at Makoshika because of a curious set of circumstances that enhanced erosion into the plains. The badlands here are on the flank of the Cedar Creek anticline, a Laramide fold that formed a bump in the horizontal rock layers. Perhaps an intense range fire destroyed the protective plant cover, and once the soft, gently tilted sedimentary rocks were exposed, they were attacked by water and rapidly eroded into narrow canyons with steep slopes. Wetting and drying of clay minerals breaks the rock up, and rain splash and surface runoff erodes rills, rivulets, and gullies. The orange rock layers are cross-bedded sandstones that are tightly cemented by iron-bearing minerals. The sandstones resist weathering, protecting the underlying, softer rock. The hard sandstone caps help to form pillars, toadstools, and tall, erosional pinnacles called hoodoos. Sandstone bridges span a few gullies, which were formed when the underlying, softer rock was eroded away. Walk the trails in the cool morning air and imagine that moment in time when life on Earth changed forever.

The Cretaceous-Paleogene boundary is a clay layer within the coal directly above the geologist's hand.

Exposed, soft sandstone and claystone are easily eroded into rills, rivulets, and gullies with each passing rainstorm.

28 Terry Badlands

Coal Beds and Clinker Caps

The Cenozoic Era dawned with life that survived the calamity at the end of the Cretaceous. The coastal plain remained largely unchanged, and deposition of the Fort Union Formation continued uninterrupted from 66 to 60 million years ago. The Terry Badlands, an erosional landscape in the Fort Union rocks north of Terry, Montana, are a hidden gem for those wanting to explore the formation's Tullock, Lebo, and Tongue River Members. The Cretaceous-Paleogene extinction boundary is not exposed, but it's a great place to see coal and clinker, a brick-red rock formed during coal seam fires. Montana was in the mid-latitudes by this time, but the climate was subtropical, with plenty of rain to grow the plants needed to eventually form coal. The sediments accumulated on a broad river plain that extended from the western mountains to the Western Interior Seaway in North Dakota, called the Cannonball Sea. The streams transitioned from braided to meandering as the gradient decreased, depositing sand in the channels and mud on the vast floodplains, which were wet and covered in lakes, swamps, and peat bogs. Plant debris accumulated in low-oxygen bogs, first forming peat. As it was buried deep by sediment, heat and pressure turned it into coal as water and methane were removed.

The coal is mined mostly in the Tongue River Member in the Powder River Basin to the south. The coal ranges from lignite to subbituminous, which are lower grades, but they became profitable because they are low in sulfur and less polluting. The Terry Badlands coal seams are very thin and

The hard clinker cap prevents the underlying, softer rocks of the Tongue River Member from eroding away, forming topography. The heat of the burning coal can melt the rock, forming unusual clinker dripstones (inset).

From Terry, travel 2.1 miles north on Highway 253 to Scenic View Road. Turn left (west), and travel 5.9 miles to Terry Badlands Overlook.

mostly lignite, a very low-grade coal that has not undergone as much heat and pressure as the higher-grade coals. The brick-red clinker formed when sedimentary rocks above and below burning coal seams were cooked at temperatures of up to 2,500°F, baking, fusing, and even melting them—rocks metamorphosed by fire. The coal fires typically started by direct lightning strikes or grass fires and burned underground for centuries. The rock name comes from the "clink" sound made when walked on or struck with a hammer, and clinker is commonly used to armor dirt roads that pass through the slick, badlands clay.

Badlands terrain is difficult to navigate due to the steep, barren hills of expansive clays and the tangle of gullies and ravines. Badlands are usually eroded into nearly flat-lying, soft sedimentary rocks with variable resistance to erosion. The clinker was essential to the formation of the Terry Badlands because it resists erosion and caps many of the hills. A dry climate with sparse vegetation is also essential to badland formation because plant cover tends to reduce erosion. Water-resistant clays also help because they shed rainwater off the slopes. Most of the erosion occurs during turbulent flash floods, producing rivulets and deepening gullies, and the eroded sediment collects in fans at the bases of the slopes. With each passing storm, the landscape changes.

The Fort Union Formation contains fossils that provide insights into how life responded after the mass extinction. Although mammals coexisted with the dinosaurs, they diversified after the extinction. The oldest primate and ungulate (hoofed animal) fossils ever found were discovered at the Terry Badlands.

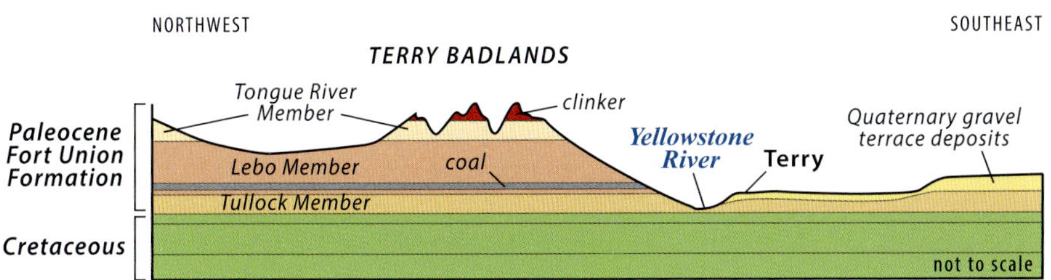

Cross section showing the members of the Fort Union Formation and clinker caps exposed at Terry Badlands. —Modified from Vuke and Colton, 2003

Colorful layers of sandstone, claystone, and black coal of the Paleocene Fort Union Formation exposed in the Terry Badlands Wilderness Study Area about 3 miles northwest of Terry. Note the red clinker at the very top of the high point.

Medicine Rocks State Park

Sandbars of a Brackish Estuary

The wonderland of arches, windows, caves, holes, pillars, and hoodoos at Medicine Rocks State Park, north of Ekalaka, are carved from a sandstone that tells a geologic story from when this place looked more like the Gulf Coast than the plains. This patch of sandstone, the Ekalaka Member of the Fort Union Formation, was deposited around 61 million years ago, some 5 million years after the nonavian dinosaurs perished. It's so different from the other deposits of the Fort Union Formation that it was designated as a new member of that formation in 2002. The deposit is interbedded with meandering stream deposits of the Lebo Member to the west and coastal marine deposits of the Cannonball Member to the east.

The quartz-rich sandstone is cross-bedded and shows that currents flowed both toward the sea, like a river flowing off the land, and toward the land, like rivers influenced by tides. Such cross-bedding is common to sandbars in modern estuaries, places where rivers transition to the sea. Mud chips in the sandstone show the river eroded clay layers from its floodplain, which was densely vegetated with subtropical plants, now preserved as fossil leaves and woody debris. Burrowing animals made underground homes typical of animals that feed on leaf litter, which is a common food source in estuaries. Predatory animals that thrived in brackish water, a mix of fresh and salty ocean water, frequented this place based on the abundance of their scales and teeth. Most belonged to animals like crocodiles and alligator gar, a hearty fish that lives on to this day, spawning in estuaries of the Gulf of Mexico. The estuaries offer protection for their young from predatory animals.

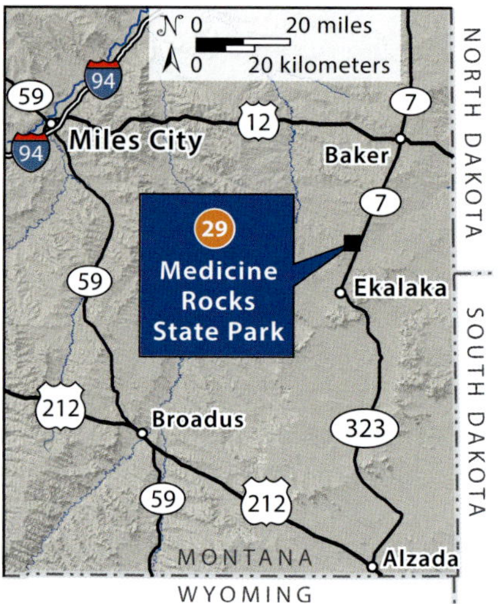

From US 12 at Baker, travel 24 miles south on MT 7, or from Ekalaka, travel 11.4 miles north on MT 7 to Medicine Rocks State Park.

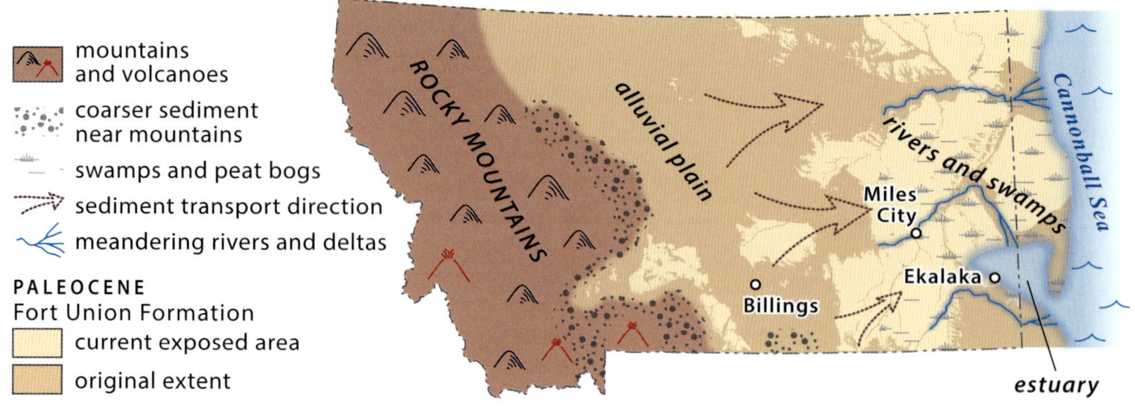

Reconstruction of the Ekalaka estuary of the Paleocene Fort Union Formation.

The rock monuments in the park were once part of a continuous sandstone deposit. Fractures have been widened by gravity, freezing and thawing, salt weathering, and wind. Large blocks of fallen rock attest to the importance of gravity. The pock marks show that alkali-rich water seeps into the porous sandstone and precipitates dissolved salts. As the salt crystallizes and expands, it pops out sand grains to form cavities. Rainwater seeping in and freezing on cold nights enlarges the cavities, along with wind blowing out the loose grains of sand. Perhaps some of the long, circular holes were fossil logs weathered out of the rocks. The rock sheds its surface so rapidly that recent human carvings into the rock become unreadable within a few years. First People have used this sacred place for religious purposes for millennia, so please be respectful and leave it as you found it.

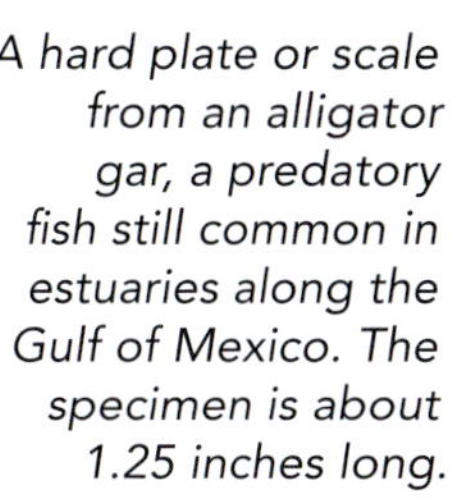

A hard plate or scale from an alligator gar, a predatory fish still common in estuaries along the Gulf of Mexico. The specimen is about 1.25 inches long.

Fossil leaves and woody debris show that the river floodplain was densely vegetated with subtropical plants.

A pillar of stream-deposited sandstone of the Paleocene Ekalaka Member of the Fort Union Formation.

30 Judith Mountains

Granitic Rocks with Unusual Quartz Crystals

We tend to think of Montana as mountainous in the west with plains to the east, but central Montana is a transitional land of island mountain ranges rising thousands of feet above the surrounding plains. Most, like the Judith Mountains near Lewistown, are domal structures cored by igneous rocks intruded largely in early Paleogene time, near the end of the Laramide mountain building. As the magma rose, it bulged up the overlying sedimentary rocks, steeply tilting them around the intrusions. The igneous rocks are rich in alkali elements like potassium and sodium, so the region is called the Central Montana alkalic province. The intrusions come in a variety of shapes, and some even erupted onto the surface, but most are laccoliths, lens-shaped intrusions where magma injected as thin sheets between sedimentary rock layers until the liquid pressure was able to raise the overlying layers into a dome. Erosion with the rising of the plains exposed the igneous rocks in the cores of many, but not all the domes.

The Judith Mountains boast more than fifteen intrusions that expose alkali-rich igneous rocks like syenite, monzonite, phonolite, trachyte, and alkali granite. For some reason, the noted alkalic rock shonkinite is not found here but is abundant to the west in the Highwood, Bears Paw, and Adel Mountains. The igneous rocks in the Judith Mountains range in age from 69 to 47 million years old, but most intruded around 50 million years ago, during Eocene time. Many intrusions were shallow, so they crystallized quickly, with large, early formed crystals floating in a fine-grained, quickly cooled matrix. Like most of the intrusions in the region, ore mineralization in the Judith Mountains sparked a gold rush. The

From Lewistown, travel 10.8 miles north to Maiden Road. Turn right (east), and travel 9.4 miles to Judith Peak Road. Turn left (north), and travel 3.6 miles to Judith Peak. The "adobe diamonds" can be found in the roadcuts near the end of the gravel road. The colors correspond to those in the adjacent diagram.

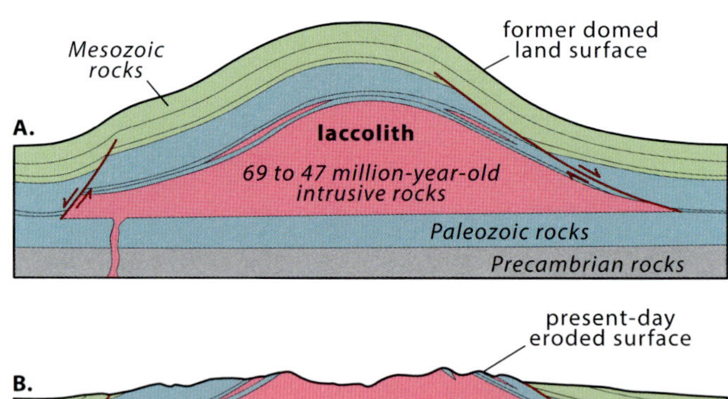

The Judith Mountains boast more than fifteen laccoliths, most of which are lens-shaped, shallow intrusions that domed the overlying sedimentary rocks.

Near the old radar station on Judith Peak Road, an alkali granite intrusion is loaded with stubby, doubly terminated smoky quartz (large, dark crystals) that weather out of the rock and can be collected off the ground. —Photo at right courtesy of Eric Eliasson

ghost town of Maiden boasted as many as 1,200 people at its peak in the 1880s.

The Judith Mountain granite is known for its smoky quartz crystals with six-sided pyramids at each end and a central band. They are called "adobe diamonds" by some. These stubby crystals formed early in the magma, allowing growth of crystal shapes before the liquid fully crystallized. Whole crystals weather out of the rock and can be collected from the granitic sand near Judith Peak. Vertical volcanic breccia pipes also formed in the region by explosions at shallow depths where magma interacted with groundwater. They did not blast from the mantle like those that contain real diamonds.

The origin of the alkalic magma is unclear. Some geologists suggest shallow subduction or crustal shortening thickened the crust, pushing the upper mantle deeper into Earth, where it melted at higher pressure. Such conditions can produce magmas unusually high in potassium and sodium relative to silica, like in the igneous rocks found here.

The Judith Mountains near Lewistown consist mostly of Eocene-age alkalic intrusions, granite-like igneous bodies rich in alkali elements like potassium and sodium.

31 Hyalite Canyon

Palisades in an Absaroka Lava Flow

While alkalic magma rich in sodium and potassium was intruding into central Montana, explosive stratovolcanoes—now eroded away—were spewing out thick piles of ash, lava, and quartz-rich volcanic debris in central Idaho, southwest Montana, and northwest Wyoming. One of the most impressive volcanic centers is the Absaroka volcanic province, extending from the Gallatin Range southeast into Wyoming. A picturesque place to see the rocks is at Palisade Falls in Hyalite Canyon south of Bozeman.

The Absaroka volcanoes were active 53 to 43 million years ago, covering nearly 9,000 square miles—an area larger than Vermont and New Hampshire combined—in up to 5,000 feet of volcanic deposits. The chemistry of the rocks suggests they were produced by subduction and partial melting of the mantle, like in the Cascades of the Pacific Northwest. However, the subduction zone was 1,000 miles to the west, so a different mechanism was the likely culprit, perhaps crustal relaxation, stretching, and decompression melting of deep, hot rocks. The viscous, silica-rich lavas were sticky, plugging vents and trapping gases. When the pressure built up, explosive eruptions released fast-moving, hot flows of gas, ash, and debris called pyroclastic flows.

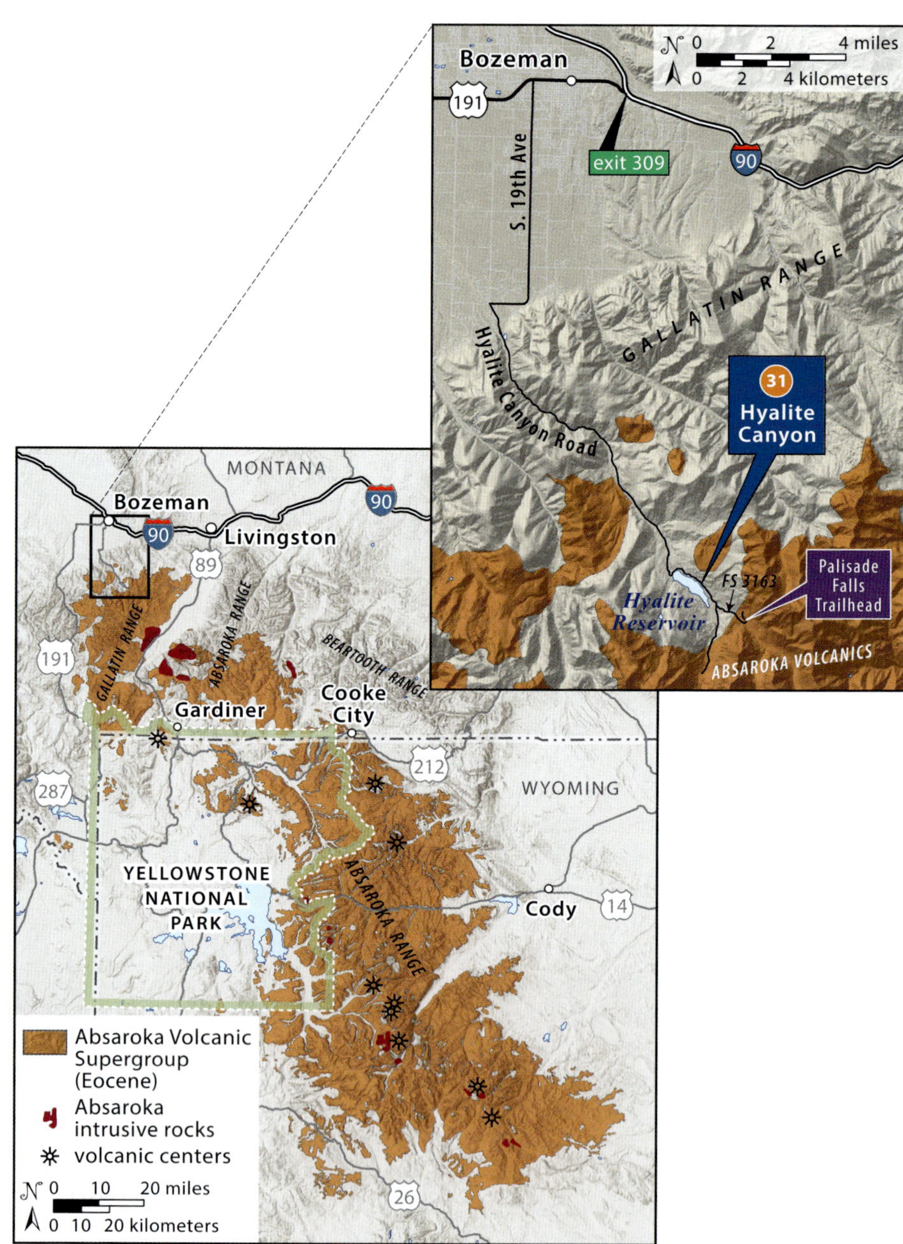

From West Main Street in Bozeman, travel 7.1 miles south on S. 19th Ave. Turn left (south) on Hyalite Canyon Road, and travel 11.7 miles to FS Road 3163. Turn left (southeast), and travel 1 mile to FS Road 3165. Turn left (north) into the Palisade Falls Trailhead parking lot, and walk the 0.5-mile hike to the falls. —Map of Absaroka Volcanic Supergroup rocks modified from Feeley, 2005

The columnar joints formed when the lava flow cooled and contracted, forming six-sided cracks.

Palisade Falls in Hyalite Canyon south of Bozeman cascades over the lip of a glacially carved hanging valley composed of columnar jointed andesite of the Hyalite Peak Volcanics.

Palisade Falls cascades over a lava flow of the Hyalite Peak Volcanics, a thick pile within the Absaroka volcanic province. In places in Hyalite Canyon, the volcanics rest on Archean basement rock exposed by Laramide uplifting and erosion. The Palisades lava flow has prominent, six-sided columns called columnar joints that formed when the lava cooled and contracted. The dark, fine-grained rock looks like basalt, but chemical analysis shows it has a lot of silica and is closer to andesite. We know the lava flowed from its vent down a stream channel because it sits on a deposit of well-rounded boulders, including a few metamorphic rocks eroded from the exposed basement.

In the old stream channel below the lava flow are lahar deposits, debris flows off the steep-sided volcanoes. These flows of water and angular rock debris were the consistency of wet concrete and inundated the stream drainages, picking up and fossilizing logs (site 32). Some lahars are inversely graded, with larger chunks concentrated at the top of the deposit. This happens when smaller particles sieve down through the rapidly moving viscous flow of mud and rock.

The origin of the waterfall is, in part, due to the hard nature of the lava rock but also to the deep erosion of Hyalite Canyon by glacial ice during the last ice age around 24,000 to 16,000 years ago. The big glacier that descended Hyalite Canyon eroded much deeper into the rock than the adjacent tributary drainages, leaving them hanging well above the base of the main canyon. The water now cascades over the lip of the hanging valley and down into Hyalite Canyon. During the lower flows of late summer and fall, the water is concentrated into a single thread that follows a crack between columns. As it descends, the water hits the tops of many columns, broadening into a veil of water that ultimately weaves its way over and around the boulders of the ancient stream channel.

Below the falls is a great exposure of bouldery stream deposits overlain by muddy debris flows (lahar deposits). The geologist's hand is at the contact.

Gallatin Petrified Forest

Fossil Trees Encased in Debris Flows

During the eruptions and outpourings of lava associated with the Absaroka volcanic province (site 31), flows of water and rock the consistency of wet concrete inundated the stream drainages. Known as lahars, these debris flows picked up entire trees, encasing them in silica-rich volcanic material when the flow came to rest. Over time, the logs were permeated with silica and fossilized as petrified wood. The Gallatin Petrified Forest, south of Livingston, is a great place to see these if you are ready for a hike. This volcanism occurred 50 million years before the Yellowstone volcanism, just to the south. Most of the petrified logs lie on their sides, but a few are standing, either having been buried in place or transported upright by viscous flows. Rocks embedded in their roots provided ballast.

The types of plant fossils entombed by the lahars show us that the Eocene climate was wet and tropical. This was a time of global warming, when temperatures were as much as 14°F warmer than today. The variety of plant fossils suggests that debris flows descending the volcanoes gathered up plants from a variety of climate zones. Spruce and fir that grew high in the mountains are mixed with hardwoods like walnut and maple from intermediate elevations. When flows came to rest in the valleys, they buried lowland species such as magnolia, avocado, and laurel. Metasequoia or dawn redwood forests were seemingly everywhere. Similar zonation occurs in the mountainous volcanic regions of central America today.

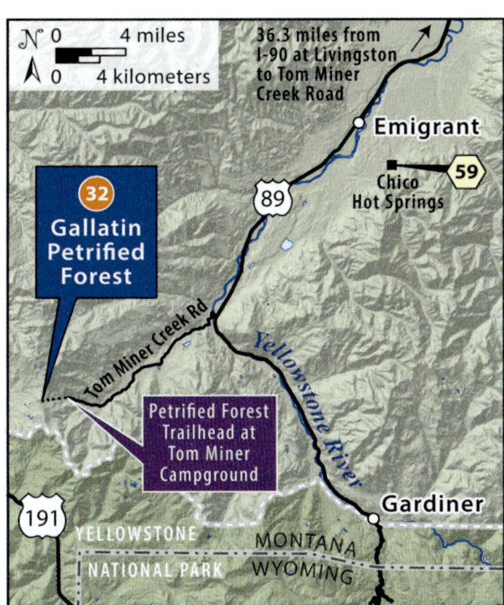

From Emigrant, travel 14.4 miles south on US 89 to Tom Miner Creek Road. Turn right, cross the Yellowstone River, and travel 11.6 miles southeast to the Tom Miner Campground. Walk the Petrified Forest Trail for several miles to the steep cliffs to see the petrified trees.

This petrified Metasequoia log is jutting out of a lahar deposit along the Gallatin Petrified Forest Trail in the Tom Miner Basin south of Livingston. —Courtesy of Chris Pace

33 Heart Mountain Detachment

The World's Largest Landslide

The Heart Mountain detachment east of Yellowstone National Park is perhaps the world's largest landslide (other than those that occur on the seafloor). About 49 million years ago, a slab of Paleozoic sedimentary and Eocene volcanic rock the size of Rhode Island catastrophically slid southeast away from the active Absaroka volcanic field near Cooke City, Montana, toward Cody, Wyoming. The slide moved at more than 200 miles per hour down a 2-degree slope, depositing debris over 1,300 square miles. It detached just above the base of the Ordovician Bighorn Dolomite, following the slope of the bedding until the landslide ramped to the surface, and momentum slid detached blocks of Paleozoic limestone across the Eocene land surface, like at Heart Mountain in Wyoming north of Cody. Shortly after the slide, additional eruptions covered much of the slide debris with volcanic deposits, helping to preserve it for us to see today.

Most of the landslide is in Wyoming, but part of the breakaway is in Montana and visible below Abiathar Peak as you look south from the northeastern entrance to Yellowstone National Park. The cliff wall exposes a steep fracture where distinctive layers of gray-colored, Paleozoic limestone are abruptly chopped off and abutted by brown-colored Absaroka volcanic rocks. The steep fracture is the place where the slide broke away from the stable rock. The abutting Absaroka volcanic rocks were likely dropped down against the Paleozoic rocks during the event and later buried by more volcanic debris after the sliding ceased. The fracture curves downward into a nearly horizontal, bedding-parallel surface of movement,

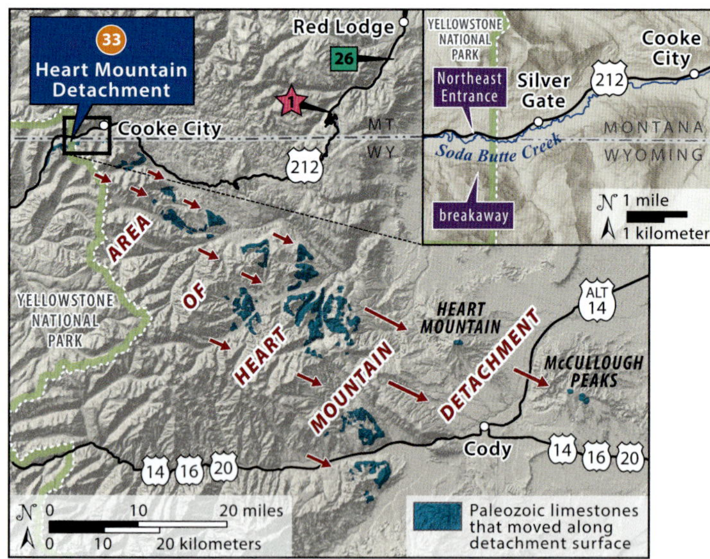

From Cooke City, travel 4.1 miles west on US 212 to the Northeast Entrance Station of Yellowstone National Park. The breakaway is visible to the south from the northeastern entrance station. Walk the short trail from the vault toilet to Soda Butte Creek for the best view. —Geologic elements of detachment modified from Losh, 2025

The Heart Mountain detachment surface, viewed here in the Jim Smith Creek drainage east of Cooke City, is located near the base of the Ordovician Bighorn Dolomite and covered by Eocene Absaroka volcanic rocks. The arrow shows the direction the slide moved. —Courtesy of Steven Losh

The Heart Mountain detachment breakaway point on the ridge between Abiathar Peak and Amphitheater Mountain, viewed to the south from the northeast entrance to Yellowstone National Park. Arrows show the direction the slide moved.

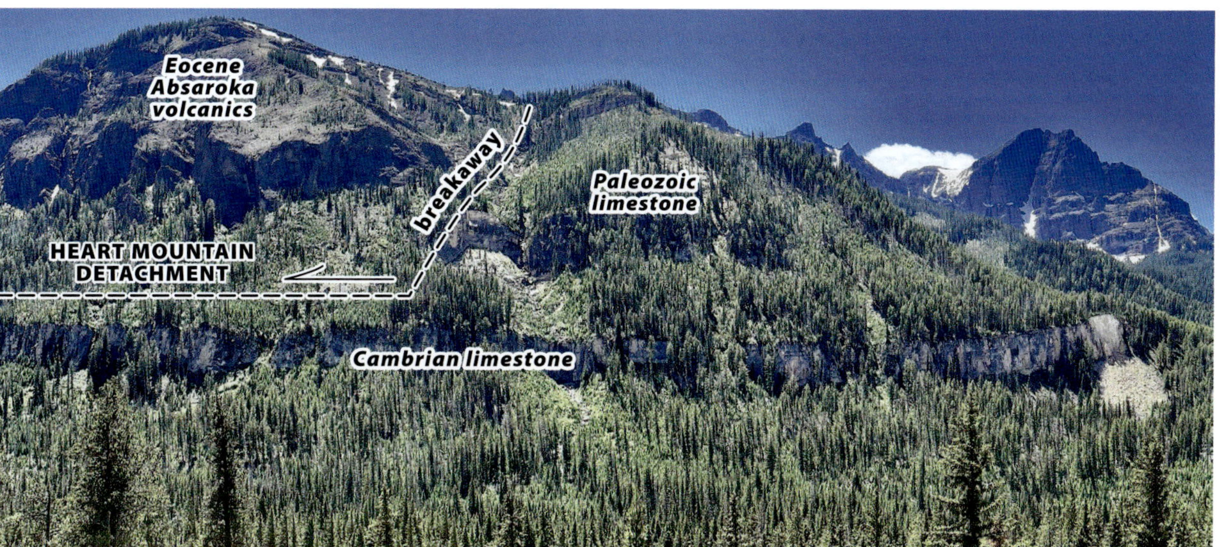

or detachment, near the base of the Bighorn Dolomite. The prominent layer of Cambrian limestone below the detachment is laterally continuous and was not involved in the landslide.

What triggered such a huge landslide, and how did it move so far and so fast over a flat surface? Landslides are common on the sides of big stratovolcanoes, such as the one that tore apart Mt. St. Helens in 1980, but this landslide moved along a deep layer in the Bighorn Dolomite, well below the volcanoes. For it to move on a gently sloping layer at depth requires a significant reduction in friction. Perhaps rising magma caused the region to dome up, creating a gentle subsurface slope, down which a large block of rock could slide. The magma might have also heated and pressurized pore water in the Bighorn Dolomite, causing the rock to lift along a single bedding plane. Add earthquake shaking, and it might start to slide. Once moving, frictional heating of the carbonate rock released carbon dioxide gas, reducing the friction even more and speeding up the landslide. We know the detachment was hot because there are veins of pseudotachylite, an unusual rock produced by frictional melting of limestone, which would have released even more gas. Friction along the detachment must have been greatly reduced because gouge or disturbed rock along the detachment is minimal. Most geologists now think the landslide took no more than 30 minutes. For information about the Wyoming portion of the landslide, see *Roadside Geology of Yellowstone Country* (2011).

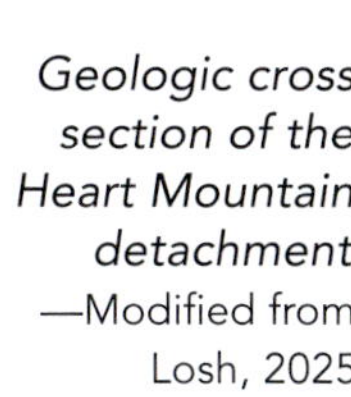

Geologic cross section of the Heart Mountain detachment. —Modified from Losh, 2025

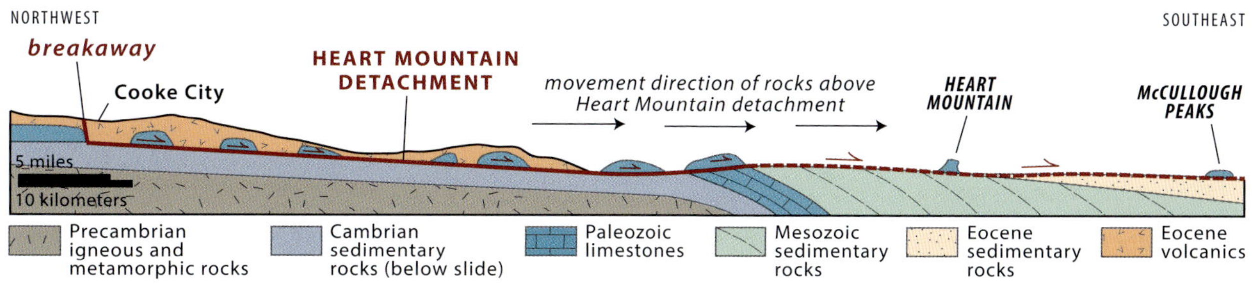

Kootenai Canyon in the Bitterroot

Mylonite and a Metamorphic Core Complex

The thickened crust formed by the Late Cretaceous plate collisions to the west began to fall apart in Eocene time. The crustal stretching formed several metamorphic core complexes, areas where domes of deeply buried rocks were brought to the surface on low-angle faults. The overlying rocks detached and slowly moved many miles away from the rising domes, thinning the crust. Several metamorphic core complexes are well documented in Montana, including one near Anaconda and the other in the Bitterroot and Sapphire Mountains.

The Bitterroot Mountains began their extensional journey around 53 million years ago as an elongate, rising dome of Belt Supergroup and granitic rocks. Granitic magma accompanied the stretching, so there was plenty of heat to soften the rocks and allow them to move eastward off the rising Bitterroot dome at great depth in Earth's crust on a gently sloping, shear zone called the Bitterroot detachment. The hot rocks of the detachment were deformed, creating a zone of streaky metamorphic rocks, called mylonite, that caps

From just north of Stevensville on US 93, turn left (west) on Kootenai Creek Road. Travel 2 miles to the Kootenai Creek Trailhead. The mylonite is well exposed on both sides of the canyon a short distance from the parking lot.

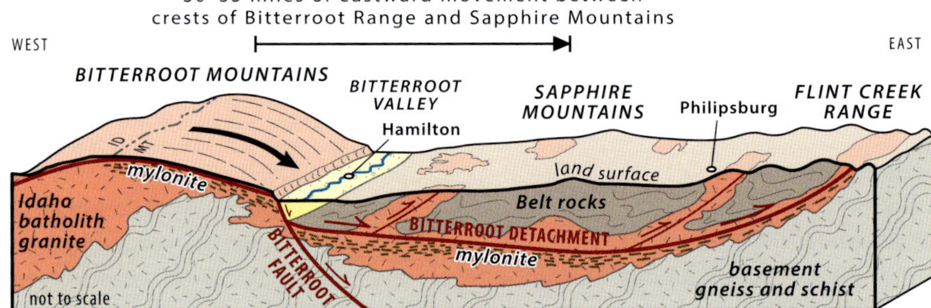

Simplified sketch showing the development of the Bitterroot metamorphic core complex and the recent offset of the detachment by active, high-angle extensional faults. —Hyndman and Thomas, 2020

The eastern flank of the Bitterroot Mountains viewed northwest from Stevensville shows the preserved smooth curvature of the Eocene Bitterroot detachment.

the highest peaks and curves down to the east at about 25 degrees. It now armors the eastern flank of the range and preserves the domal shape of the old detachment. The less metamorphosed rocks that detached from the dome now reside far to the east in the Sapphire Mountains. The Bitterroot Valley between the dome and the detached block is filled with thousands of feet of sediment eroded from the adjacent mountains.

Most of the canyons eroded into the eastern flank of the Bitterroot Mountains expose the mylonite, but the Kootenai Creek Trail, west of Stevensville, displays important features just a short walk from the parking area. The Bitterroot mylonite is more than 1,000 feet thick in places and consists of slabby rocks that look as though they moved like a sliding deck of cards. On rock surfaces, linear features created by stretched minerals point to the east, the direction the overlying rocks moved along the detachment fault. The rocks of the Bitterroot Mountains were pulled out from under the overlying rocks of the Sapphire Mountains as they moved east.

The change from compression to extension in Eocene time is not well understood. Some attribute it to crustal relaxation as the subduction zone moved west, while others argue northward sliding of landmasses along the west coast wrenched the crust apart. Whatever the cause, movement on the Bitterroot detachment ceased 30 million years ago, but the mountains continued to rise along high-angle extensional faults. The Bitterroot fault, at the base of the mountains, offsets glacial deposits that are only 14,000 years old and can produce large earthquakes. You can see the recent high-angle faulting where the eastern slope of the range front abruptly steepens near the base of the mountains.

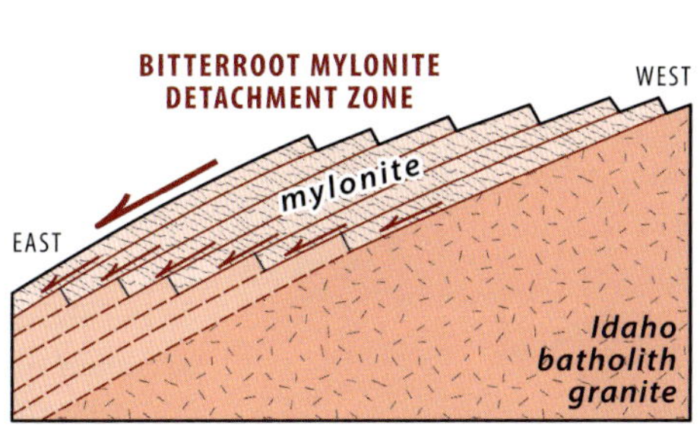

Simplified sketch of the internal layering of the Bitterroot mylonite detachment zone showing down to the east movement like a sliding deck of cards. —Hyndman and Thomas, 2020

Along the Kootenai Creek Trail just past the trailhead on the north canyon wall is an asymmetrically deformed rock in the mylonite of the detachment zone. The arrows show top-to-the-east sense of shear, which is the direction of movement of rocks detached from the Bitterroot Dome.

The layering within the detachment zone along the south wall of Kootenai Creek suggests the rocks moved eastward (direction of the arrows) like a sliding deck of cards.

Medicine Lodge Beds

Fossiliferous Lake Sediment in a Down-Dropped Valley

From 48 to 20 million years ago, the crust continued stretching apart, raising mountains and dropping valleys along a north-south rift zone that overlapped the trend of the Sevier fold-and-thrust belt. Maybe the thick pile of compressed rock collapsed, or the subducting plate progressively sagged into the mantle, causing hot mantle to well up. Regardless, many valleys formed and filled with alluvial fan, stream, and lake sediments of the Renova Formation. These rocks preserve the bones, teeth, and poop (coprolites) of mammals as they diversified. Sharing the land were insects, fish, amphibians, reptiles, and plants that are recognizable to us today.

The Renova Formation tends to form light-colored deposits that some consider dirt, but it's endlessly fascinating. Lake sediments spanning the Eocene-Oligocene boundary can be found in an abandoned Gilmore & Pittsburgh Railroad cut near Grant, west of Clark Canyon Reservoir. Called the

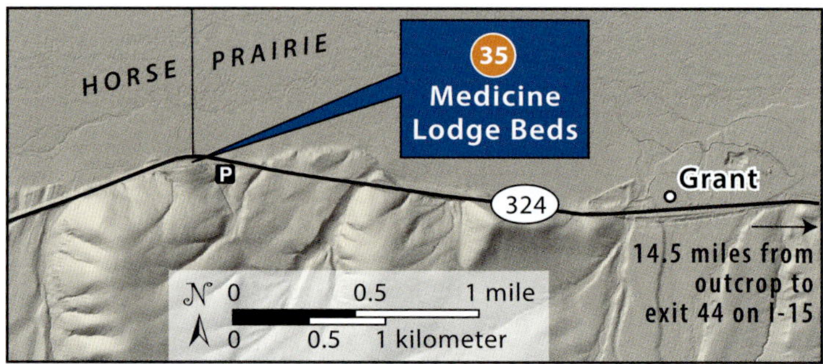

From Grant, travel 2 miles west on MT 324. Park in the pullout to the left (south), and walk west into the abandoned railroad cut to collect fossil plants.

Lake deposits of the Eocene to Oligocene Medicine Lodge Beds of the Renova Formation are well exposed in an abandoned railroad cut west of Grant.

Metasequoia or dawn redwood is the most common plant fossil found on the bedding surfaces of the lake siltstones.

Gypsum crystals cut across sedimentary layering, showing they likely precipitated in fractures in solid rock rather than by evaporation of lake water during deposition.

Medicine Lodge Beds, the sediments accumulated in a rapidly dropping valley that was flanked by rising, fault-bounded mountains. Sand and gravel were deposited on alluvial fans and in streams draining the mountains. The streams built deltas into the lake, burying peat bogs. The burial converted the peat to low-grade coal used by the railroad. The Eocene climate started out tropical and steamy but got progressively drier with time, with grassy savannas expanding and grazing animals evolving to live there.

The sediments in the railroad cut consist mostly of parallel, thin layers of ashy claystone, siltstone, and fine-grained sandstone. Such fine-grained sediments tend to accumulate where waves and currents are subdued, allowing the grains to settle out of the water, like in lakes. The subtle changes in grain sizes from one layer to another reflect changing stream flows bringing sediment into the lake, which occurs during flooding or seasonal variations. We can be confident these are lake deposits because they contain scales and bones of freshwater fish and countless tiny shells of ostracods, a crustacean common in lakes. Visitors to the site are typically drawn to the shiny pieces of gypsum littering the ground. These fibrous crystals look interbedded with the other deposits, as though precipitated during evaporation of the lake, but they also cut across layers, so they likely precipitated from groundwater into fractures in the rock.

This site is particularly rich in fossil plants, mostly grasses, leaves, needles, and carbonized wood, that grew near the lake or were brought in from the surrounding highlands. One of the mountain-dwelling plants that is well represented was the dawn redwood, a deciduous conifer related to sequoias that sheds its needle-like leaves. Fossil plants are excellent indicators of past climate, and the Medicine Lodge Beds span a major climatic change and associated mass extinction marked by the Eocene-Oligocene boundary, around 33.9 million years ago. The early Eocene was a time of greenhouse warming, with global average temperatures up to 14°F warmer than today, but by the end of the epoch, a reduction in greenhouse gases and the formation of the Antarctic Circumpolar Current caused glaciation, cooling, and global extinction. The plant fossils in the Medicine Lodge Beds record this event, showing a marked decline in diversity and a pronounced change from subtropical to cool and dry-adapted temperate plants.

36 Black Butte

Basalt atop the Youthful Gravelly Range

You can reach Black Butte on good gravel roads, and it is visible from anywhere atop the Gravelly Range. From Alder, travel 29.1 miles south on Upper Ruby Road, and stay slightly left onto Lazyman Creek Road for 4 miles. Turn left (east) on Short Creek Road for 3.6 miles, and turn right (south) on the Gravelly Range Road for 6.5 miles to the base of Black Butte.

In Eocene time, volcanic fields in Montana were still spewing mostly silica-rich lavas, but by Oligocene time, the composition of the igneous rocks became more basaltic. The reasons for the transition continue to be debated, but basaltic magma typically forms from crustal stretching and decompression melting of the mantle. A picturesque remnant of an ancient basaltic volcano that was active about 23 million years ago occurs atop the Gravelly Range in southwest Montana. Black Butte is a plug of dark-colored basalt that crystallized in the volcanic vent, a now-prominent peak (10,547 feet elevation) that rises more than 1,200 feet above the surrounding terrain.

Basalt lava is relatively low in silica and rich in iron and magnesium. It tends to form where pressure is released on the hot rocks of the mantle, allowing the chemical bonds in the minerals to break and form magma. Different minerals melt at different temperatures, so when the mantle partially melts, it forms a magma rich in calcium plagioclase and pyroxene but with less olivine than the mantle rock from which it was derived. The basaltic magma rises and injects into fractures formed by the stretching crust and erupts hot, fluid lava onto the surface.

The basaltic igneous rocks in the Gravelly Range intruded and erupted in a variety of forms, including as dikes, sills, lava flows, and also plugs that fill volcanic vents. The basaltic plug of Black Butte has small, green crystals of olivine that formed below the surface and were brought up during the eruption of the volcano. Erupted lava flows of similar age in the area tend to be loaded with vesicles or holes formed as gas was escaping from the top of the flow as it rapidly solidified into rock. These fluid lavas flowed downhill into stream

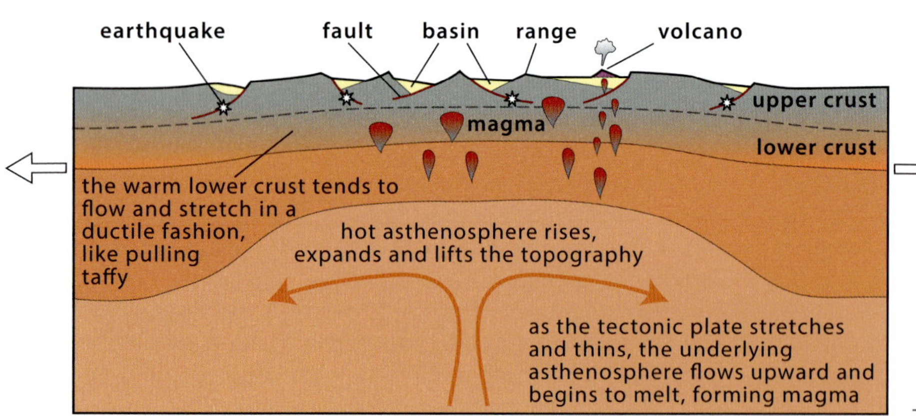

Extension results in crustal thinning, pressure release, and partial melting of the mantle to form basaltic magma.
—Modified from the National Park Service

channels and are interbedded with stream and lake sediments. They now form the topographic high-points atop the range, like on Lion Mountain southeast of Black Butte, because the hard basaltic rocks resist erosion more than the surrounding sedimentary rocks. What was once the topographic low is now the topographic high, a geomorphic feature called inverted topography.

The lava flows provide sobering insight into how much the Gravelly Range has been uplifted since the volcanism occurred. When the Black Butte volcano was active, the Gravelly Range could not have existed because lava flows from that time are interbedded with stream and lake deposits of the Renova Formation, sediments that accumulated in the lowlands of the Renova Basin. The mountains likely started rising during renewed crustal stretching around 17 million years ago. The nearby Huckleberry Ridge Tuff, a ground-hugging pyroclastic flow erupted from the Yellowstone volcano 2.1 million years ago (site 40), has been uplifted 3,600 feet since that eruptive event. This world-class uplift rate shows the landscape is very young and rapidly changing.

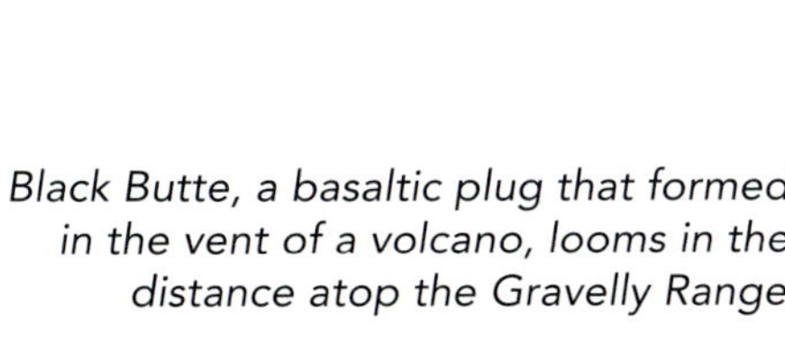

Black Butte, a basaltic plug that formed in the vent of a volcano, looms in the distance atop the Gravelly Range.

37 Timber Hill near Dillon

Not All Ash Falls from the Sky

In 1805, Lewis and Clark followed the broad valleys of the upper Missouri River in search of a northwest passage to the Pacific Ocean. About 17 million years earlier, that route and the river didn't exist. The thick crust of ancient North America intersected the Yellowstone hot spot near the present junction of Oregon, Idaho, and Nevada around 17 million years ago. Extension had begun throughout the Intermountain West, and the hot spot added to the mix as magma domed, stretched, and cracked the crust in a radial pattern over the thermal bulge (see figure on page 12). In western Montana, existing extensional faults were rejuvenated, while new northeast-trending faults dissected the once stable Renova Basin. From 17 to 4 million years, the ancestral Missouri River passed through the basin on its way to Hudson Bay, depositing ashy river and alluvial fan sediments of the Sixmile Creek Formation.

Many white layers of ash are visible in the mountain ranges of southwest Montana, like those in the Sweetwater Range at Timber Hill east of Dillon. We think of ash as falling from the sky, but these deposits got here when torrents of water, ash, and pumice were unleashed down the ancestral Missouri River during mega-eruptions of the Yellowstone volcano, perhaps when the water of a caldera-filling lake was released. Each flood traveled hundreds of miles from its source, spreading out across the valley floor and dumping up to 100 feet of ashy debris. Some came in pulses, ripping up large blocks of ash and incorporating them into the next flood of water and ash. Streams reworked the ash, and when the flood was over, grasses left behind root casts in the newly formed soils. Between floods, the valley filled with rounded stream gravels and alluvial fan deposits from the nearby mountains.

There are about a dozen ash beds, so the ancestral Missouri River captured flood deposits from many of the hot spot eruptions as the North American plate moved slowly to the southwest over the stationary plume. Stream gravels in the Sixmile Creek Formation increase in size with time because the river gradient increased as the hot spot's thermal bulge encroached closer to the Timber Hill area. Around 6 million years ago, basaltic lava erupted from the Heise volcanic field near Rexburg, Idaho, and flowed here down the old drainage. The hard basalt has been topographically inverted, forming three distinct mesas, including the top of Timber Hill.

The Sixmile Creek Formation is now exposed in many of the northwest-trending mountain ranges in southwest Montana, including the Sweetwater Range. These mountains

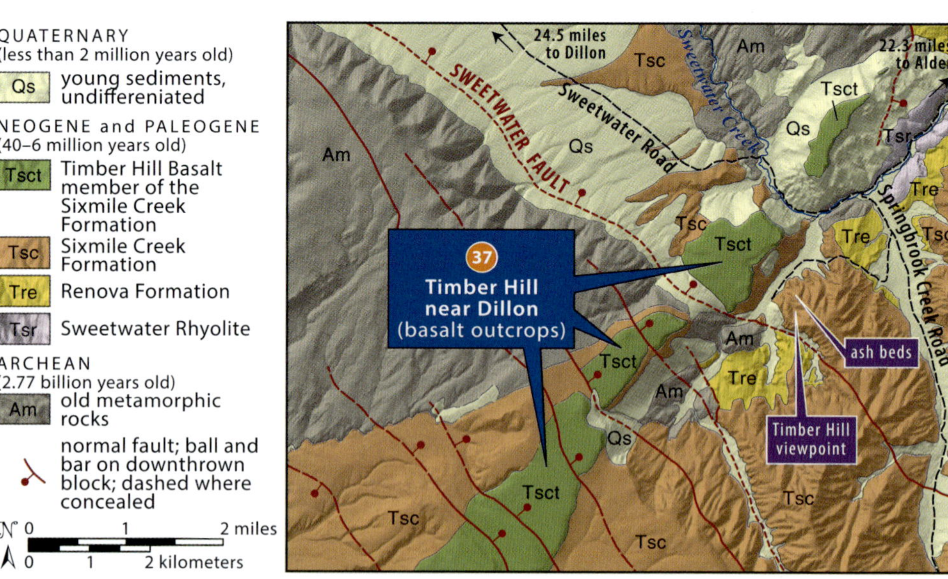

From Dillon, travel 24.5 miles east on the Sweetwater Road. Turn right (east) on Springbrook Creek Road, and stay right on an unnamed dirt road at about 0.6 miles. In 1.5 miles, thick ash beds are on the left (east), then stop in 0.5 miles at the Timber Hill viewpoint.
—Modified from Mosolf, 2023

This thick ash deposit was transported northeast by viscous floods from Idaho during a caldera-forming eruption of the Yellowstone hot spot around 10 million years ago. The ash here rests on soil (peach-colored layer) formed on the river's floodplain.

Once the ashy flood events were over, grasses recolonized the ground surface leaving behind fossil root casts called rhizoliths.

could not have existed when the ancestral Missouri River was active because rivers don't flow uphill and over mountains. The mountains must have started forming after the Timber Hill basalt lava flowed north from Idaho and are, therefore, as young as 4 million years. The new mountains formed as the hot spot arrived at its present position in northwestern Wyoming, and the thermally raised terrain left behind collapsed along normal faults into northwest-trending basins and ranges that rearranged the older drainage basins. In a normal fault, the overlying block slides down a sloping surface, creating a basin, while the underlying fault block is uplifted, creating a mountain range. The northwest-trending faults are active and capable of producing large magnitude earthquakes, like the magnitude 7.3 earthquake on August 17, 1959, that triggered a landslide, buried a campground, and began impounding Quake Lake (site 58).

Cross-sectional view looking northwest at the Sweetwater fault, an active, extensional normal fault that cut the Missouri River stream deposits (Sixmile Creek Formation) and Timber Hill basalt that were transported northeast from the track of the Yellowstone hot spot in Idaho.

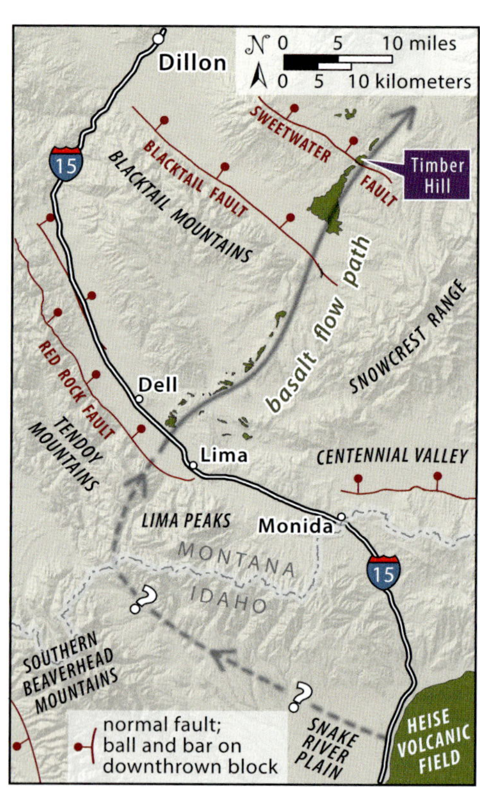

The Heise volcanic field vent and flow path for the Timber Hill basalt flow around 6 million years ago, before the existing northwest-trending mountain ranges, including the Sweetwater Range, existed.
—Modified from Mosolf, 2023

Mission Range

Basin and Range Topography in Northwest Montana

Montana was named for the Spanish word *Montaña*, which describes the mountainous terrain of the western part of the state, yet the countryside is equally characterized by large valleys. In northwest Montana, the valleys are dropping down along north-trending normal faults that are active and hazardous. These down-dropped valleys are part of the Basin and Range province, a region of extension-induced, north-south-oriented mountains and valleys that extends in the United States from eastern California to central Utah, and from Arizona to Montana.

The most stunning range front in Montana might well be the Mission Range near St. Ignatius. Towering more than 6,000 feet above the valley floor, the steep range front is an ominous sign of recent uplift and the potential for big earthquakes. The Mission fault at its base dips steeply west toward the basin and becomes flatter at depth. The rocks exposed in the Mission Range are Mesoproterozoic Belt Supergroup sedimentary rocks. In the adjacent valley, the Belt rocks are tilted down to the east into the fault, buried under a few thousand feet of sediment. The Belt rocks reemerge on the upslope side in the hills of the Bison Range to the west. The normal fault and its basin likely started dropping in Eocene time, like others in the region, and the Belt Supergroup rocks appear to have been offset by more than 16,000 feet over that long history.

The Mission Range from the Confederated Salish and Kootenai Tribe Bison Range near Charlo.

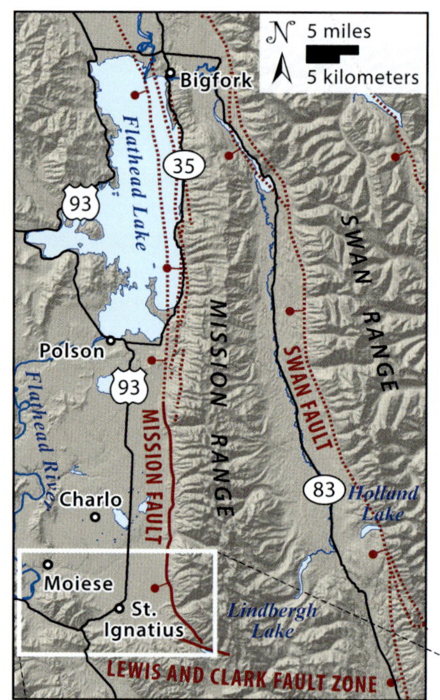

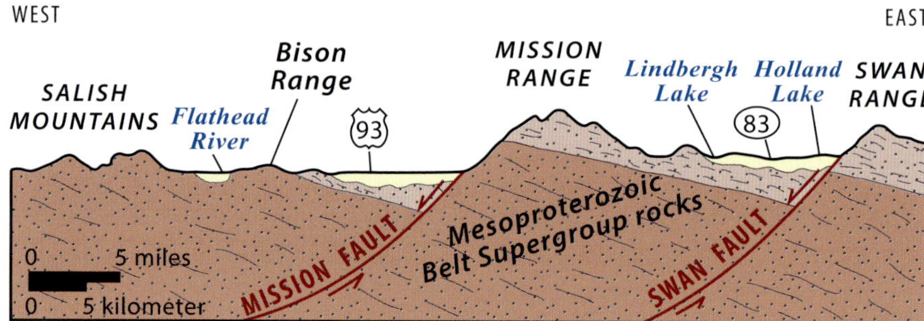

Cross section across the Mission and Swan Ranges north of St. Ignatius. —Hyndman and Thomas, 2020

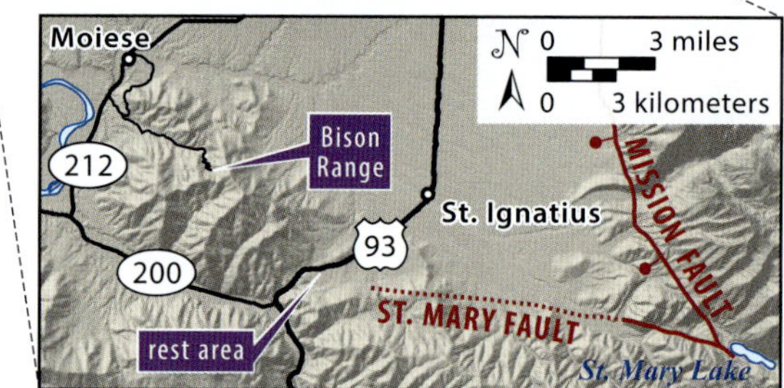

A good view of the Missions can be had from the hilltop rest area on US 93 south of St. Ignatius or from the roads in the Confederated Salish and Kootenai Tribe Bison Range. From MT 212 near Moiese, turn east onto Bison Range Road. Stop at the visitor center or continue onto Red Sleep Mountain Drive for great views. The Mission fault extends from the north end of Flathead Lake to St. Mary Lake south of St. Ignatius. Solid lines mark the known location of the faults, while the dotted lines show the trace of the fault buried under surficial deposits. —Faults from Stickney and others, 2000

The steep mountain front of the Mission Range as seen from the Confederated Salish and Kootenai Tribe Bison Range near Charlo.

Today, the Mission fault is within a seismically active zone called the Intermountain Seismic Belt, an area of mostly normal faults of the Basin and Range province from Nevada through British Columbia. The Mission fault starts near Big Fork in the north as multiple faults, including several mapped under the waters of Flathead Lake. They merge into one fault extending south to St. Ignatius, where the range reaches its greatest elevation and is chopped off by the St. Mary fault of the Lewis and Clark fault zone. There is good evidence at the surface and beneath the waters of Flathead Lake that the faults are active, including breaks in the ground surface, called fault scarps, that formed during large magnitude earthquakes. Bedrock facets are also present—steep, smooth surfaces of the fault plane formed from recent movement.

Although there have been no major earthquakes on the Mission fault in recent times, geologists discovered a 20-foot vertical offset in glacial deposits along the front. The rupture happened 7,700 years ago and perhaps caused a very large (greater than 7.0 magnitude) earthquake. The recurrence interval (the frequency of similar earthquakes) was calculated to be between 4,000 and 8,000 years, so another big earthquake is overdue. If it were to happen today, the shaking might cause the collapse of water-saturated ground, landslides, or sloshing of Flathead Lake, called a seiche. A surface rupture of a fault under Flathead Lake could generate a giant wave that washes over homes and businesses along the shoreline. If the ground were to significantly drop during such a quake, the eastern shoreline could be permanently submerged.

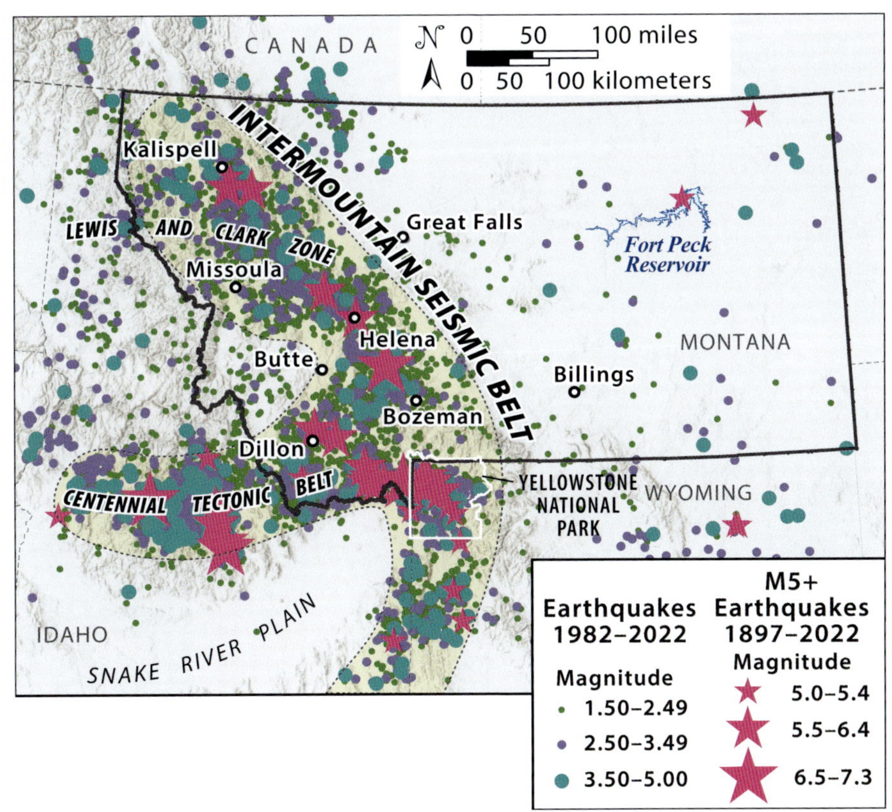

Map of the Intermountain Seismic Belt in Montana showing earthquakes greater than magnitude 1.5 between 1982 and 2022 and historic earthquakes greater than magnitude 5.0 (pink stars). —Courtesy Mike Stickney, Montana Bureau of Mines and Geology

Snowcrest Range Benches

Former Valley Floor Surfaces

The river valleys in southwest Montana are often flanked by raised flat surfaces, called benches, that gently rise toward the adjacent mountains. Some extend all the way to the mountain front, while others are cut off and isolated by erosion. Most are truncated before reaching the valley bottoms due to erosion by the modern rivers. Some benches are pediments, gently sloping, erosional surfaces with thin caps of gravel that formed when the climate was arid during Pliocene time, from about 5.3 to 2.5 million years ago. Pediments tend to occur where stream canyons emerge from the mountains but can coalesce to form nearly continuous benches that gently slope (less than 7 degrees) downward from the mountain front.

Once you start looking, benches seem to be everywhere in Montana, but not all are pediments. Some are alluvial fans, like the Cedar Creek fan near Ennis, built during the last ice age and now eroding. Some are abandoned stream channel gravels that are now elevated above the modern drainage, like those of Rock Creek near Red Lodge, and many are stream terraces, former floodplains left behind as the rivers erode downward, like the striking terraces of the Madison River (site 48) below Quake Lake. So, how do you know you are looking at a pediment? Let's check out the benches extending from the Snowcrest Range adjacent to the East Fork of Blacktail Deer Creek southeast of Dillon.

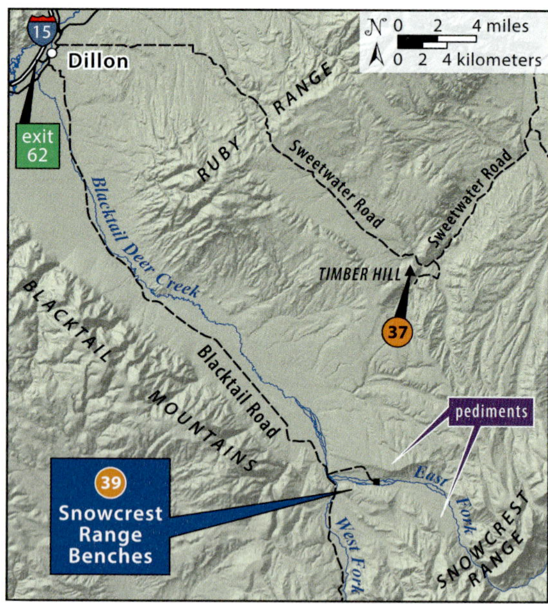

From Dillon, travel 27.2 miles southeast on the Blacktail Road. Turn left (northeast) on the East Fork Blacktail Road, and travel 2.6 miles to an old ranch next to the creek, all part of the Blacktail Wildlife Management Area. The top of the pediment can be reached by walking from the ranch.

The pediment is capped by about 20 feet of stream gravel (foreground) eroded from the Snowcrest Range (in the distance).

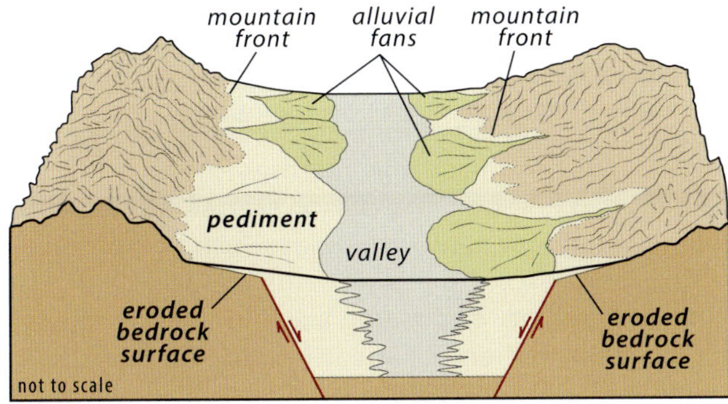

Pediments tend to form along the flanks of mountain ranges in desert climates because sheet flooding erodes back mountain fronts and leaves behind thin deposits of stream gravel on the eroded bedrock surface.

mechanisms include deep weathering of the bedrock surface, which makes it is easy to erode, coupled with episodic stream transport of sediments. In arid environments, the lack of vegetation reduces soil formation and stream channel stability, so water spreads out into broad sheets of rapidly flowing water called sheetfloods. These highly erosive flows are like sandpaper eroding a broad swath of the underlying bedrock, leaving behind a thin veneer of gravel as the flow wanes. Over time, mountains erode back or upslope as the pediment forms. During Pleistocene time, the pediments in southwest Montana were deeply eroded as the land rose and the climate became wetter, deepening the valleys and creating substantially more topographic relief with the mountains.

The bedrock below the bench tops are soft sedimentary rocks of the Miocene Sixmile Creek Formation that are at an angle to the slope of the bench and deeply eroded. The eroded surface is overlain by a thin (about 20 feet thick), sheet-like deposit of rounded gravel composed mostly of rock types found upslope in the Snowcrest Range but now disconnected from it by erosion. These are hallmarks of a pediment, and the benches on either side of the East Fork of Blacktail Deer Creek were once connected as a continuous pediment formed by water emanating from the Snowcrest Range. A Pliocene age is likely because the surface formed on late Miocene deposits and was deeply eroded during Pleistocene time.

The processes of pediment formation have been debated for more than a century. It is now recognized that they occur in a variety of climates, tectonic settings, and rock types. Most proposed

A Pliocene pediment surface slopes gently downward from the Snowcrest Range to the east (in the distance). This former valley floor surface was mostly eroded during the wetter climates of Pleistocene time, and now the large valley (at left) is occupied by the undersized East Fork of Blacktail Deer Creek. The dashed line shows the approximate contact between the eroded Sixmile Creek Formation below and the overlying gravels above.

Yellowstone's Huckleberry Ridge Tuff

One of the World's Largest Volcanic Eruptions

Yellowstone National Park, home to the largest, most violent volcano in the world, sits atop a plume of heat rising from the mantle known as the Yellowstone hot spot. The North American plate has moved southwest over the stationary hot spot for millions of years, arriving at its present location about 2 million years ago. Since the hot spot's arrival beneath Wyoming, the Yellowstone Plateau volcanic field has experienced three mega-eruptions that, combined, spewed out more than 1,000 cubic miles of volcanic debris (tephra) and formed giant collapse structures called calderas. The most recent mega-eruption, 639,000 years ago, ejected so much material from its magma chamber that a caldera formed measuring 45 miles long by 30 miles wide. The magma chamber below is of similar size and less than 10 miles deep. Two smaller fingers of magma reach to within 3 miles of the surface, forming two cracked and faulted domes within the caldera. The granitic magma is gas-charged, viscous, and only about 9 percent liquid, so it's very explosive.

Caldera-forming eruptions occur when rising magma cracks the overlying rock and releases pressure at the top of the magma chamber. The dissolved gas rapidly expands, shattering the magma into fine, glassy ash and pumice. The gas and tephra explode out of vents at supersonic speeds into the stratosphere, depositing ash and tuff from pyroclastic flows. The ground then collapses into the void at the top of the magma chamber, creating a caldera. Once formed, the caldera fills up with sticky rhyolitic lava, so the most recent Yellowstone caldera is difficult to see because it has been accumulating thick flows of rhyolite for the last 600,000 years.

Although Montana was peripheral to the calderas, the landscape was altered by the thermal doming and eruptions. The pyroclastic flows—fast-moving, ground-hugging clouds of hot gas, ash, and pumice (tephra)—reached into Montana and solidified into a rock called tuff. The most expansive deposit is the Huckleberry Ridge Tuff, expelled during the collapse of the Island Park caldera 2.1 million years ago.

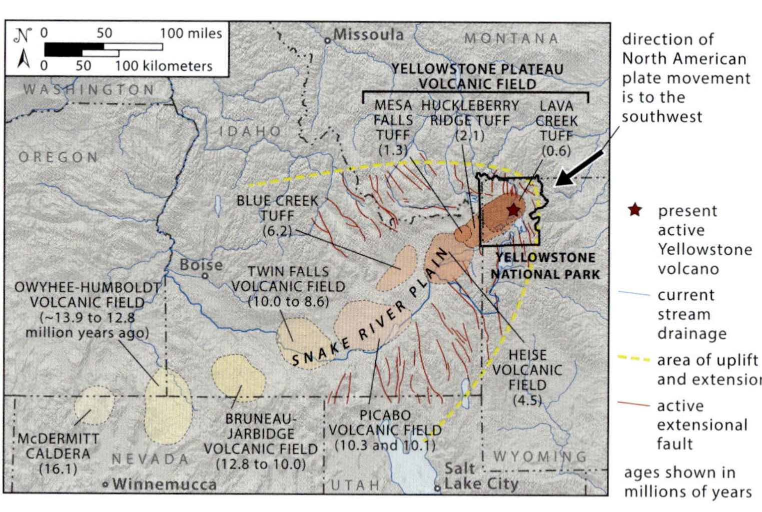

Map showing the track of the Yellowstone hot spot and the three caldera-forming eruptions of the Yellowstone Plateau volcanic field. —Modified from Pierce and Morgan, 1992

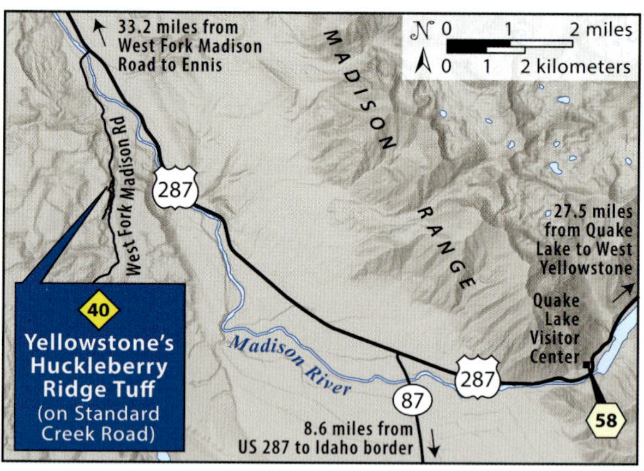

From Ennis, travel 33.2 miles south on US 287. Turn right (southwest) on the West Fork Madison Road, cross the Madison River, and travel 2.3 miles to the split in the road. Take the right (west) fork (Standard Creek Road) for 0.2 miles, park at the first curve, and walk up the road to view exposures of the Huckleberry Ridge Tuff (44.8685, −111.5825).

Built in 1903 from local columnar basalt, the Roosevelt Arch in Gardiner is the formal gateway to Yellowstone.

Pyroclastic flows tend to follow topographic lows, so this one flowed northward for tens of miles into Montana's Yellowstone, Gallatin, Madison, and Red Rock River valleys. In the Madison Valley, the flow reached more than 50 miles north of the caldera, where the tuff is exposed along the West Fork of the Madison River and easily reached on Standard Creek Road. The eruption that produced the Huckleberry Ridge Tuff was at least 2,500 times larger than the eruption of Mt. St. Helens in 1980 and perhaps only matched in recent times by the Toba eruption in Sumatra 74,000 years ago.

To say the Madison Valley had a "terrible, horrible, no good, very bad day" (as the title of a famous children's book pronounces) during that eruption is an understatement. At first, huge volumes of ash fell from the sky. Once the upward momentum of the erupted debris was overcome by gravity and the density of the tephra, some of it rapidly

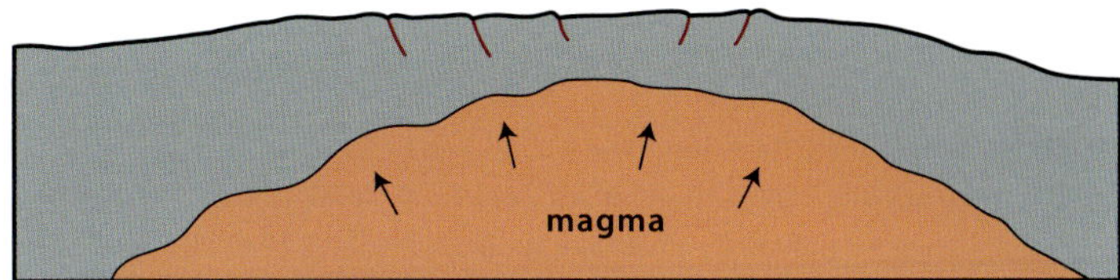

A. Magma expands and pushes upward

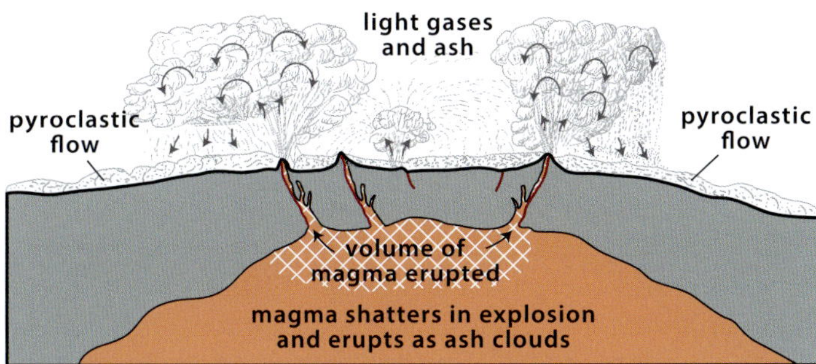

B. Erupted superhot ash and gas falls; pyroclastic flows can travel at hundreds of miles per hour for many miles

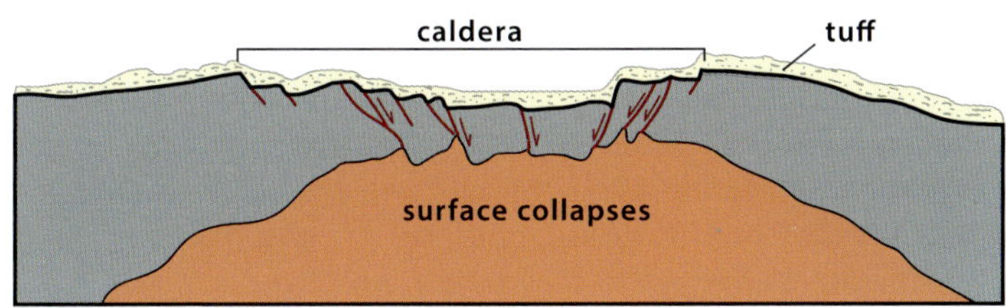

C. Hard, dense rock forms as ash grains stick together

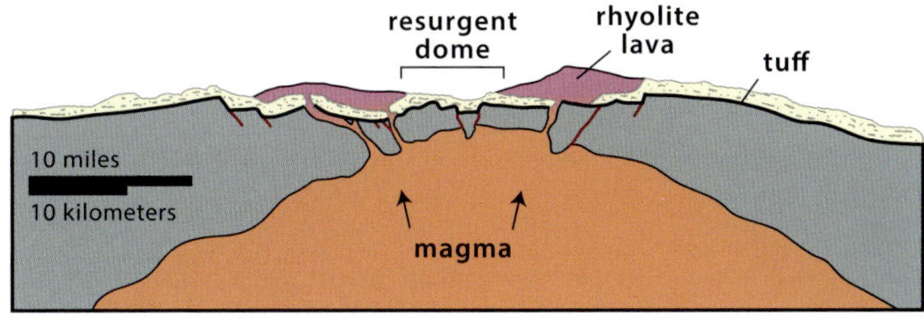

D. Lava lifts roof, oozes out cracks, and fills caldera

Diagram illustrating the sequence of events that formed the Yellowstone caldera 639,000 years ago. Welded tuff forms when there is enough heat in a pyroclastic flow that the ash and pumice within it are squashed and fused together. —Modified from Keefer, 1971; Lowenstern and Hurwitz, 2008; Fritz and Thomas, 2011

collapsed to the ground, squirting out and flowing down the valley at speeds of hundreds of miles per hour and at temperatures as high as 1,800°F. It was literally a fiery hell on Earth, and everything in its path was incinerated. One or two more pulses followed, piling up some 300 feet of ash and tephra that thins northward to around 3 feet about 15 miles south of the town of Cameron.

The exposure above the West Fork of the Madison River displays many different elements of the Huckleberry Ridge Tuff. Below the tuff is ash, likely deposited as airfall prior to the pyroclastic flow. The ash grades into a hard, fine-grained rock with curved fractures that is likely baked ash. It's abruptly overlain by several feet of vitrophyre, a black obsidian with white feldspar crystals. It formed when the pyroclastic flow hit the ground and cooled so quickly that crystals could not form. The visible feldspar crystals formed slowly in the magma chamber prior to the eruption and then were entrained in the fiery flow. The tuff above the vitrophyre was deposited by two pyroclastic pulses, each marked by different minerals and textures and likely separated in time by years to decades. Most of the tephra was hot and pliable when it settled because the ashy groundmass is fused and the pumice is squashed, forming a very hard, dense rock called welded tuff.

People commonly ask if Yellowstone will erupt again soon. In the last 2.1 million years, caldera-forming eruptions happened about every 700,000 years, most recently 639,000 years ago. But, before you panic, that interval is based on only three eruptions and so does not make for a robust statistical argument that an eruption is imminent. The likelihood of the Yellowstone volcano producing a caldera-forming eruption in any given year is estimated by the US Geological Survey at only 0.001 percent. In contrast, there is a nearly 2 percent chance of dying while driving the busy, winding roads of Yellowstone National Park. More worrisome are big earthquakes on normal faults that are triggered, in part, by thermal doming around the hot spot. One such fault is visible from the Palisades River Access on the Madison River (44.9962, −111.6596), where it cuts the Huckleberry Ridge Tuff with several hundred feet of offset.

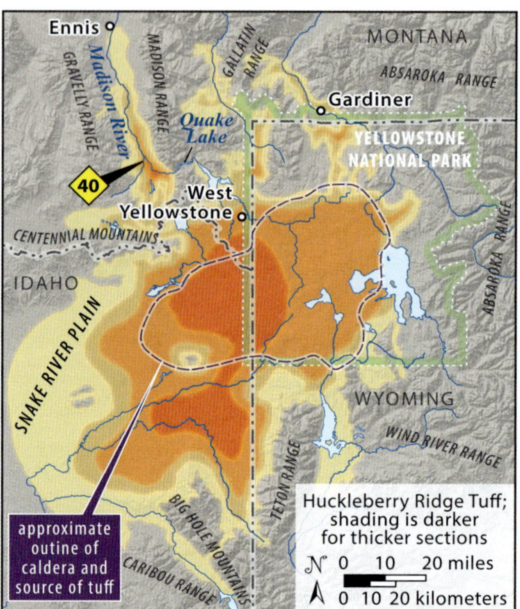

Maximum distribution of the combined members of the Huckleberry Ridge Tuff. —Modified from Christiansen, 2001

Airfall ash with parallel lamination formed as the ash rained from the sky.

Excellent outcrop of the 2.1-million-year-old Huckleberry Ridge Tuff on USFS Road 237 (Standard Creek Road) above the West Fork of the Madison River south of Cameron. The exposure displays all the elements of a pyroclastic flow deposit (44.8685, −111.5825).

Welded tuff with distinctly squashed pumice called fiamme.

Vitrophyre with white feldspar crystals.

Baked ash showing curved fractures from the fused glass.

41 Shonkin Sag

Outburst Floods from Glacial Lake Great Falls

During the Pleistocene ice ages, a massive continental glacier called the Laurentide ice sheet formed in Canada and flowed south. It advanced into Montana at least twice, first during the Illinoian glaciation from 190,000 to 130,000 years ago and again during the Late Wisconsin glaciation, from 35,000 to 11,000 years ago. Although the second ice advance was not as big, it destroyed most of the evidence of the older glaciation. The ice was no more than 1,000 feet thick in northern Montana, but it still altered the landscape as it eroded rock, deposited till, impounded lakes, and changed the path of the Missouri River, which had flowed north into Canada's Hudson Bay prior to glaciation.

When the ice blocked the Missouri and other north-flowing rivers, they were diverted east along the southern edge of the ice, eroding out big channels or sags that are now dry or occupied by much smaller streams or lakes. Because the land surface sloped down to the north, river and glacial meltwater ponded against the ice, creating at least six glacial lakes connected by rivers flowing eastward across the plains. Some 20,000 years ago, you could have paddled a canoe along the edge of the ice from Cut Bank to Glendive. The largest lake was Glacial Lake Great Falls, formed when the ice flowed against the northern edge of the Highwood Mountains, blocking the ancestral Missouri River. The lake was 600 feet deep at Great Falls, but its unstable ice dam likely failed many times, creating outburst floods that eroded a now-abandoned channel called Shonkin Sag northwest of Geraldine.

The Shonkin Sag flood story is less well-known than the Glacial Lake Missoula floods but ranks as one of the most important geological sites in the world. Although we know precious little about the outburst flooding, like how often it happened, we do know the water plunged 300 feet over a half-mile-wide precipice at Lost Lake, forming a waterfall nearly twice as high as Niagara Falls. To top it off, the channel exposed a rare igneous rock called shonkinite, injected from the Shonkin Sag laccolith as dark fingers into the light-colored Cretaceous Eagle Sandstone in Eocene time.

We also know that the outburst floods stopped at least 11,000 years ago as the ice sheet retreated to the north. The Missouri River reclaimed its preglacial path from Great Falls to Coal Banks

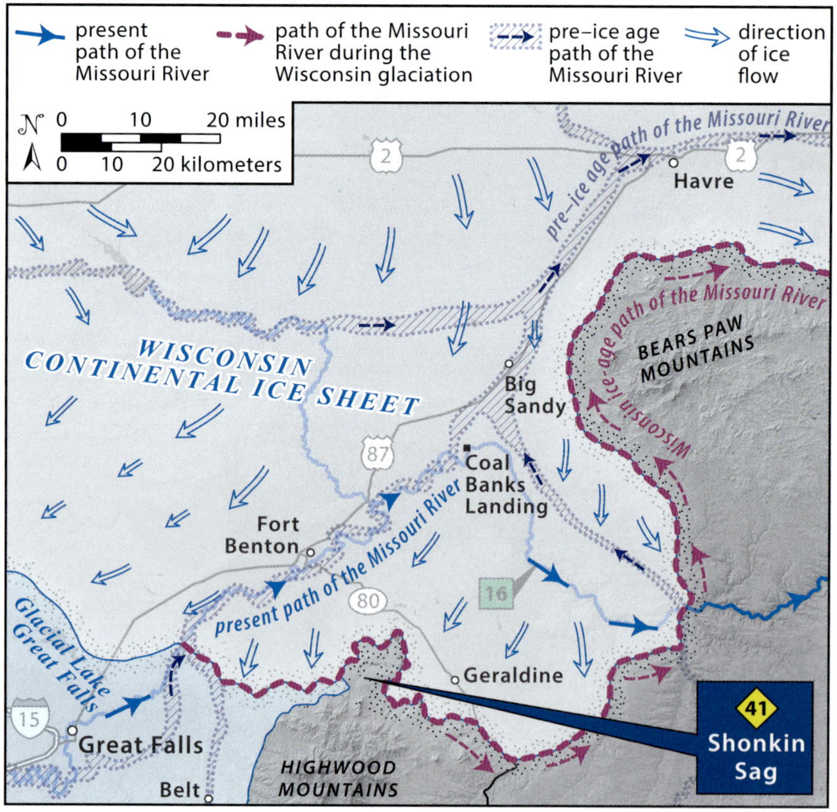

The Pleistocene continental ice sheet blocked the path of the Missouri River, diverting it east along the ice margin and impounding glacial lakes, like Glacial Lake Great Falls. —Modified from Fullerton and others, 2012

Looking upstream at the phantom waterfall or cataract at Lost Lake near Geraldine that was carved from outburst flooding each time an ice dam on Glacial Lake Great Falls failed. Access by permission only. —Courtesy Travis Mahn

Landing but then turned east because glacial debris filled its former channel to the north. Glacial Lake Great Falls must have been a sight to behold in its time. The ice sheet formed high cliffs of blue ice along its northern shore that calved off icebergs into the water. They drifted across the lake, dropping boulders plucked from Canada as they melted. Those scattered boulders, along with shoreline benches and pink-colored lake sediments, are all that remain of the lake. Fossils from this time show that horses, antelope, elk, camels, bison, ground sloths, musk oxen, mastodons, and mammoths all thrived on the tundra south of the ice. Their presence, in turn, supported predators like the American lion, dire and gray wolves, giant bears, and saber-toothed cats with 7-inch canines. Humans, who were in Montana by at least 14,000 years ago, witnessed this ice-age wonderland. Unfortunately, there is no public access or viewpoint for modern-day visitors.

Logan Pass in Glacier National Park

Ice Age Sculpting of the Mountains

Most of the mountains of Montana owe their rugged nature to alpine glaciation during the Pleistocene Epoch. Geologists once agreed that there were four great ice advances separated by interglacial episodes in which most of the land ice melted, but more recent deep-sea core data show that there were perhaps as many as twenty. In Montana, we see clear evidence of just two major episodes of glaciation, both late in Pleistocene time. In the mountains, the older advance is called the Bull Lake glaciation, similar in age to the Illinoian glaciation on the plains. The younger Pinedale glaciation coincides with the Late Wisconsin glaciation of the plains and peaked around 20,000 years ago.

The most iconic place for alpine glacial features is Logan Pass in Glacier National Park, which was covered by an ice cap 20,000 years ago. The ice flowed downhill, plucking and abrading the underlying bedrock along the way. Previously rounded peaks were attacked on all sides, making gnarled, rocky pinnacles called horns, like Reynolds Mountain. Where glaciers eroded broad ridges on two sides, the intervening rock became narrow, serrated arêtes, like the ridge west of Clements Mountain. Ice plunging down the peaks gouged out deep scoops in the rocks, leaving bowl-shaped depressions called cirques, many of which now hold lakes called tarns, like Hidden Lake. As the glaciers descended the valleys, they scoured the old V-shaped valleys into broad U-shapes like St. Mary River valley. Smaller tributary glaciers could not keep pace with the erosion, so they were left hanging, some forming waterfalls like Bird Woman Falls.

The glaciers ultimately reached down valleys to where they melted as fast as they moved forward and no longer advanced. Rock debris (till) accumulated at the end and sides of the glaciers into ridge-like piles called moraines. Where glacial meltwater was dammed by moraines, the valleys flooded, like at Lake McDonald. Where meltwater streams were able to breach the moraines, the torrents of water made outwash plains of sand and gravel.

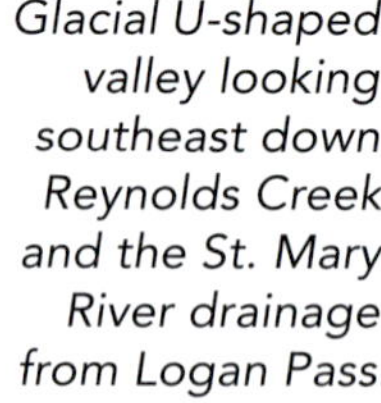

Glacial U-shaped valley looking southeast down Reynolds Creek and the St. Mary River drainage from Logan Pass.

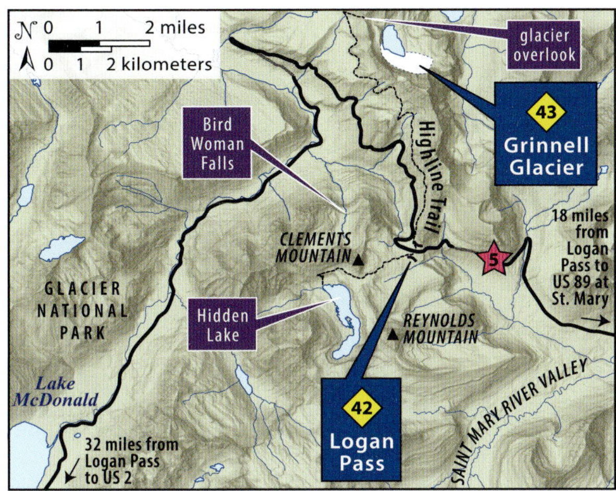

From St. Mary, travel 18 miles west on the Going-to-the-Sun Road to Logan Pass parking lot. Walk the 2.7-mile roundtrip trail to Hidden Lake Overlook.

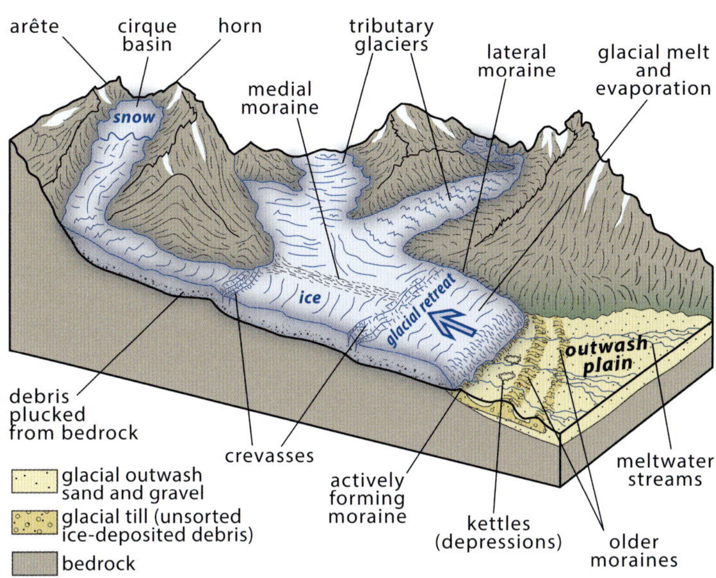

The typical erosional and depositional features formed by alpine glaciation. —Hyndman and Thomas, 2020

This pile of glacial till is a moraine once deposited by a glacier (a stagnant remnant of the ice is still there but hidden from view) below Clements Mountain on the Hidden Lake Trail, Logan Pass. The inset shows the unsorted nature of till with its variety of grain sizes.

View of glacially sculpted mountains from the Hidden Lake Overlook, a 1.4-mile-hike from the Logan Pass parking lot in Glacier National Park. A glacier once occupied this valley. You might also see a bighorn sheep, grizzly bear, or mountain goat.

43

Grinnell Glacier

A Remnant from the Little Ice Age

See map on page 105.

Glaciers still exist in Glacier National Park, and while you can't easily see any close-up at Logan Pass, Grinnell Glacier is just 3 miles north of the pass as the crow flies. While one of the bigger glaciers in the park today, it is much smaller than it used to be. Near the end of Pleistocene time, a milder climate caused glaciers to retreat around the globe. Between about 10,000 and 6,000 years ago, during the Holocene Climate Optimum, it warmed enough that nearly all glaciers in Montana completely melted, including Grinnell Glacier. The climate cooled a few degrees from 1300 to 1870, during the Little Ice Age, causing surviving glaciers to grow a bit and new, small glaciers to form, like Grinnell Glacier, which dates to 500 years old. All glaciers started melting rapidly by the end of the twentieth century from human-induced climate change. Of the estimated 150 glaciers in Glacier National Park in 1850, only 25 small, shrunken glaciers remain today. Grinnell Glacier has lost nearly half its footprint in the last 50 years, and researchers estimate that most of the glaciers will be gone from the park by the end of this century. So, go see the glaciers while you can, and also use your new knowledge to identify the glacial features of the past, and begin to imagine the icy landscape of the ice ages. Grinnell Glacier can be reached from Many Glacier Hotel by a challenging, 10.6-mile roundtrip hike (1,600 feet of elevation gain). The boat shuttle across Swiftcurrent Lake and Lake Josephine cuts the roundtrip hike to 7.4 miles. The Grinnell Glacier Overlook can be reached from Logan Pass on the Highline Trail, but it is 15 miles roundtrip.

Grinnell Glacier in 1938 (left) and in 2021 (right). Note how the glacier occupied the entire cirque in 1938 (where the lake is today). —Courtesy of T. J. Hileman and Glacier National Park

Glaciation in the Flathead

Moraine Surfing South of Polson

The Flathead Country in northwest Montana is sold to tourists as "glacier country" because it is the gateway to Glacier National Park, but the moniker also rings true to geologists because the valley was invaded by several thousand feet of glacial ice in Pleistocene time. The ice advanced into the Flathead and Mission Valleys at least twice from the Cordilleran ice sheet in Canada and the ice cap over the northern Rocky Mountains. The bodies of ice merged at the northern end of the Mission Range to form a continuous ice surface, called the Flathead lobe, that reached deep into the Mission Valley and deposited two moraines.

The southernmost one, near Ninepipes and known as the Mission moraine, could be as old as the Bull Lake glaciation, around 140,000 years. The peaks at the northern end of the Mission Range are more rounded than those to the south because the ice overrode and ground them down during the Bull Lake advance. The flat surface north of St. Ignatius is an outwash plain from meltwater streams draining the southern end of the Bull Lake ice. Some of the circular pothole lakes on the Mission moraine are kettles made when blocks of buried glacial ice melted, and the ground collapsed. Others are pingos, depressions formed when freezing and thawing heaved the ground into mounds that collapsed when their ice cores melted.

Starting around 30,000 years ago, the Flathead Lobe advanced again, reaching some 2,500 feet thick near Kalispell during the last glacial maximum around 20,000 years ago during the Pinedale glaciation. It stopped around Polson, where the ice sat long enough to build the huge Polson moraine, which impounds Flathead Lake today. Every now

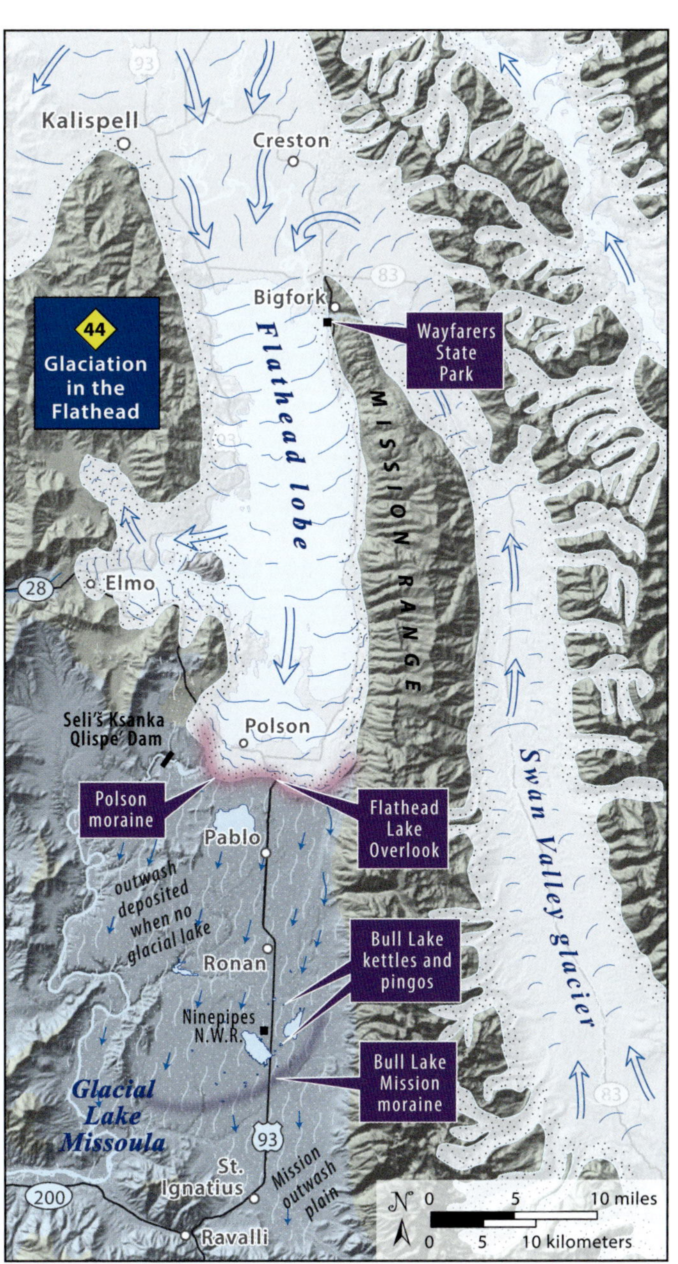

From St. Ignatius, travel 8.7 miles north on US 93 to the Ninepipes Lodge area. The rise in topography just past Post Creek is the terminus of the Mission moraine. Notice the circular kettle or pingo lakes atop the moraine. Continue 16.5 miles north to Flathead Lake Overlook Scenic Turnout, which is on the Polson moraine and has a great view of Flathead Lake to the north. The Pinedale ice advanced into the Flathead Valley around 20,000 years ago. —Ice location modified from Smith and others, 2020

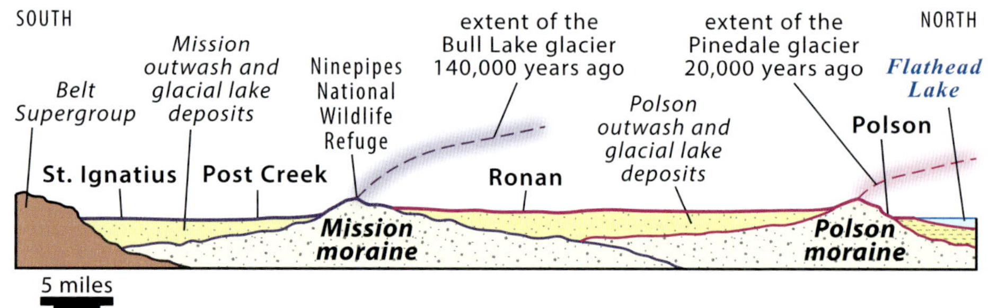

Profile of the present Mission Valley between St. Ignatius and Polson showing the location of the moraines and outwash plains. Glacial Lake Missoula sediments likely cover all deposits south of the Polson moraine. —Modified from Hyndman and Thomas, 2020

Rocks embedded in the bottom of a glacier flowing south (the direction of the hammer handle) scratched these gouges into Belt Supergroup rocks at Wayfarers State Park in Big Fork.

The circular ponds visible from US 93 near the Ninepipes Reservoir may be glacial kettles or pingos. The glaciated Mission Range rises to the east.

View to the north of Flathead Lake from the Flathead Lake Overlook Scenic Turnout on US 93, which is located on the Polson moraine.

and then during the ice age, Glacial Lake Missoula (site 46) lapped against the south end of the Flathead lobe, creating a complex moraine built of ice-deposited rock debris and subaqueous alluvial fans. The lake also cut shorelines into the hills around the valley, and many of the big boulders scattered about were dropped by icebergs floating in the lake. When Glacial Lake Missoula was absent, meltwater breached the moraine and shed outwash gravel southward into the Mission Valley. The moraine and outwash stand nearly 1,000 feet above the level of Flathead Lake, showing the great thickness of the ice at its terminus. As the ice began retreating around 13,000 years ago, meltwater was impounded behind the moraine, forming a body of water called Glacial Lake Flathead. As the ice retreated, the lake grew until it overflowed the moraine, cutting a channel down to hard bedrock (near Seli'š Ksanka Qlispe' Dam). The bedrock prevented the lake from draining even further.

Today's Flathead Lake fills a 370-foot-deep-hole, much deeper than the rest of the Flathead Valley north of the Polson moraine. Perhaps the depression was formed by the weight of the glacier, or maybe it formed when stagnant ice in the glacial debris melted. Bedrock along the shore of the lake displays gouges and scratches made by rocks embedded in the bottom of the glacier as it flowed south.

45 Eureka Drumlins

Streamlined Hills Formed by Flowing Ice

In addition to the Polson and Mission moraines (site 44), the ice sheet left behind other evidence of its passing. From the Canadian border south to Eureka, the valley is covered with elongated, whale-shaped hills called drumlins. They are composed of glacial till that was deposited beneath the flowing ice. After it was deposited, the southeast-flowing ice sheet molded the glacial debris into streamlined hills parallel to the flow. The blunt end of the hills face into the flow, and they taper downstream. The drumlins formed near the end of the glacier as the ice thinned because thicker ice would have been much too heavy and spread out the till. They may have formed here because there was sufficient meltwater at the base of the ice to lubricate and speed up the flow enough to sculpt and erode the till into streamlined shapes. People are now building structures on the steep-sided drumlins, which poses a hazard because the unconsolidated deposits are prone to landslides.

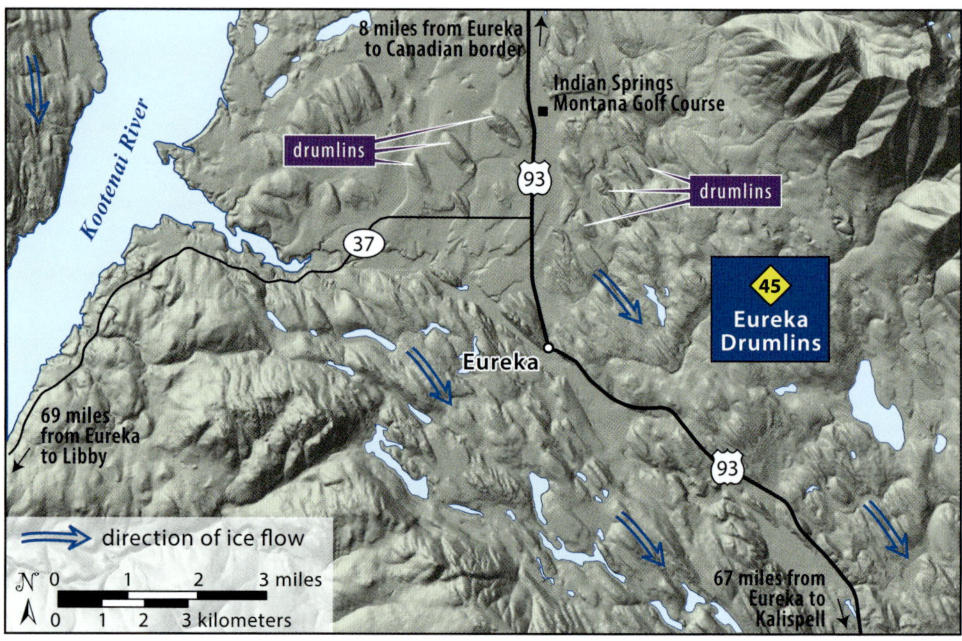

The Eureka drumlin field shows up well on the shaded relief map, and the drumlin hills are visible from the Indian Springs Montana Golf Course along US 93 north of Eureka.

Huge drumlins, shaped under the glacial ice, are clearly visible from US 93 at the Indian Springs Montana Golf Course north of Eureka. The ice flowed in the direction that the hill tapers (see arrow).

Glacial Lake Missoula

Floods of Biblical Proportions

The story of Glacial Lake Missoula and its humongous floods has caught the imagination of the public, but be advised that some renditions of the story are not based on science. Good data show us that during the Pinedale glaciation, from 20,000 to 14,000 years ago, the Purcell Trench lobe of the Cordilleran ice sheet advanced into northern Idaho and blocked the Clark Fork River. At its peak, the ice dam was more than 2,100 feet high, forming a lake that flooded 3,000 square miles of western Montana with more than 500 cubic miles of water, about the size of Lake Ontario. Periodically, the rising water caused the ice dam to float and break, rapidly draining the lake. The glacier then readvanced and created a new lake that drained when the dam failed yet again. A variety of data show us the lake filled and spilled at least forty times, sending torrents of water that scoured the Channeled Scabland of eastern Washington en route to the Pacific Ocean and dumped huge, ice-rafted boulders everywhere in between.

The most visible evidence of the lake in Montana is the horizontal, wave-cut shorelines on the steep hills above Missoula, each formed by wind-driven lake waves lapping against the hills. The high-water mark is 4,200 feet, so from the oval at the University of Montana at about 3,200 feet, you can look up and imagine being submerged under 1,000 feet of water. Over time, each lake was a bit smaller and lasted for fewer years than the one before as the climate warmed and the ice dam thinned. The highest shoreline benches are older than

The horizontal lines on Mt. Jumbo in Missoula are wave-cut shoreline benches of Glacial Lake Missoula.

At Missoula, notice the horizontal, wave-cut shorelines on the treeless, western faces of Mt. Jumbo and Mt. Sentinel (which sport the letters L and M, respectively). Find the light-colored, varved lake sediments along US 93 about 6 miles north of Arlee. To see the giant ripples at Camas Prairie, take MT 382 north from Perma for 13.1 miles, and stop at the Montana Department of Transportation interpretive sign.

those below, and the lowest bench was cut the last time the lake grew large enough to reach Missoula. By about 14,000 years ago, it was over.

When the ice dam floated, Glacial Lake Missoula emptied in a matter of days, releasing huge glacial lake outburst floods into Washington. In Montana, draining lake water reached 80 miles per hour through narrow gorges, stripping away soils and leaving deposits that could only have been made by fast-moving water, like the giant ripples at Camas Prairie near Hot Springs. Lake water poured over several passes, excavating deep funnels into the bedrock and spreading torrents of water and gravel across Camas Prairie. What remained behind is amazing: giant asymmetric ripples, some thousands of feet long and 45 feet high! The ripples are composed of cobbles and boulders. Many floods crossed Camas Prairie over time, but the last one to pass through

Glacial Lake Missoula sediments are exposed at a pullout a few miles north of Arlee on US 93.

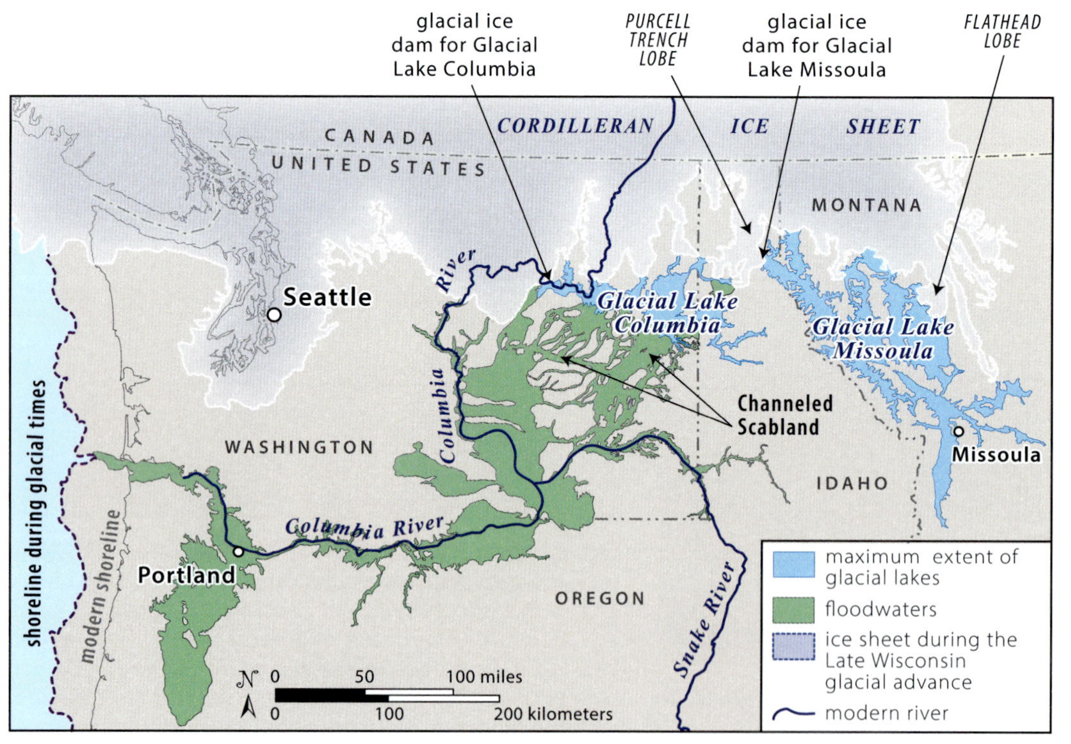

Map of Glacial Lake Missoula and the Channeled Scabland during the last major advance of glacial ice from Canada. —Modified from the Ice Age Floods Institute

Aerial view to the south of the giant ripples in Camas Prairie from an outburst flood of Glacial Lake Missoula (note the farm building at the bottom for scale). —Courtesy Dave Bennett

here must be the one we see today because older ripples would have been stripped away by the next outburst flood.

In the early years, high velocity draining of the lake left behind gravel, but as the lake got smaller, fine-grained lake deposits survived the drainings. Nicely exposed lake deposits north of Arlee on US 93 have repeated silt-to-clay layers or varves that record a year of sediment accumulation in the lake. During the summer months, the melting ice transported finely ground glacial sediment into the lake, giving the water a milky-turquoise color. The sunlit surface waters teemed with microscopic plants called diatoms, but little else survived in the silty, frigid, unstable lake. The melting stopped during the winter months, cutting off the sediment supply and causing the diatoms to perish as the surface waters froze. All winter long, diatoms and clay slowly rained to the bottom, building a thin layer of dark-colored sediment that contrasts with the light-colored layers of summer.

Red Rock Lakes in the Centennial Valley

Active Faulting and Generations of Lakes

Off the beaten path in southwest Montana is the Centennial Valley, a basin actively dropping down along the Centennial fault as the ground collapses with the passage of the Yellowstone hot spot. The valley and adjacent Centennial Mountains are young, having mostly formed since the eruption of the Huckleberry Ridge Tuff (site 40) from a hot spot caldera some 2.1 million years ago. Small areas of the tuff occur atop the range as well as in the valley, suggesting that the range has been uplifted thousands of feet on the Centennial fault since the tuff's eruption. The fault has also broke the ground surface of modern alluvial fans, with the uplifted side of the fault more than 30 feet higher than the down-dropped side, indicating the normal fault is active. It ruptures every 3,000 years, producing large earthquakes. The valley hosts Red Rock Lakes National Wildlife Refuge, established in 1935 to protect trumpeter swans, which were on the brink of extinction in the lower 48 states at the time.

The Red Rock Lakes formed in the last 10,000 years or less as alluvial fans from the Centennial Mountains built out into the valley and dammed the Red Rock River, the most distant headwaters of the Missouri River. The tilt of the valley down into the Centennial fault caused the river and now the lakes to be trapped up against the range front. The lakes are only about 6.5 feet deep, mostly due to the limitations of the alluvial fan dams.

This is not the first time the Centennial Valley has boasted a lake. During Pinedale glaciation, the ancestral Red Rock River drained in the opposite direction, flowing northeast into the Madison River through an old canyon now marked by a chain of steep-walled lakes, including Elk, Hidden, Cliff,

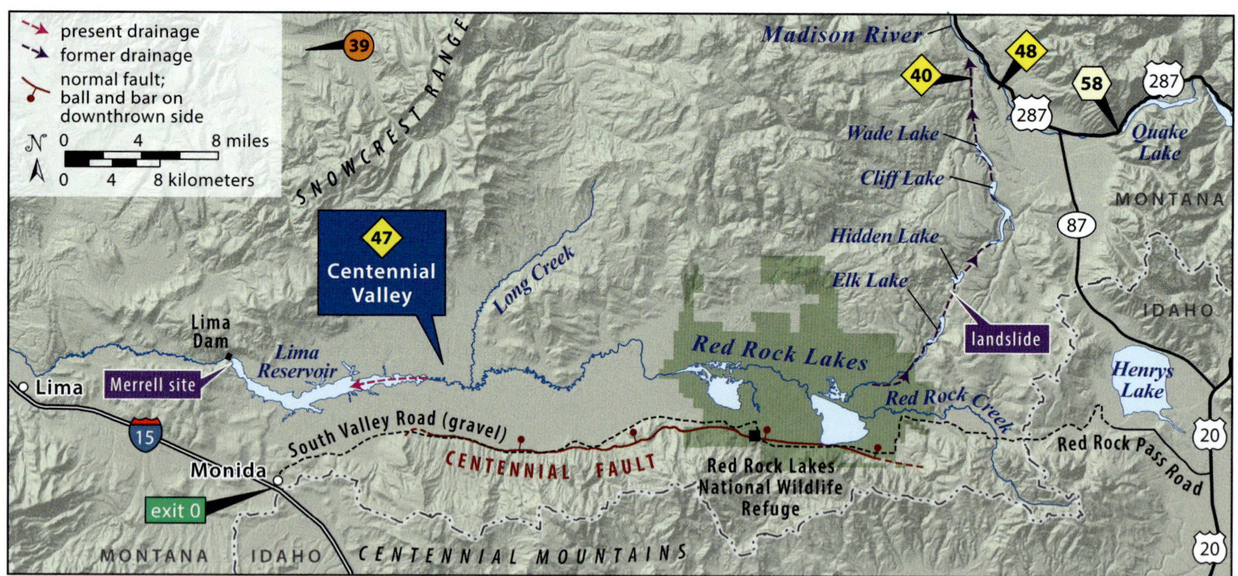

Centennial Valley drained to the northeast into the Madison Valley along the chain of Elk, Hidden, Cliff, and Wade Lakes until a landslide at the northeast end of Elk Lake dammed the drainage during Pinedale glaciation, creating Pleistocene Lake Centennial and forcing drainage to the west. The Red Rock Lakes National Wildlife Refuge headquarters is 27.4 miles east of Monida on South Valley Road.

and Wade Lakes. Sometime between 30,000 and 20,000 years ago, a landslide at the northeast end of Elk Lake dammed the old drainage and created Pleistocene Lake Centennial. The lake rose high enough to release water over a topographically low area at the western end of the valley near the present-day Lima Dam, forming the current path of the Red Rock River. At its highest level, the lake was about 65 feet deep, as shown by the old, wave-cut shoreline benches visible on the north side of the valley near Long Creek. Fine-grained, layered lake sediments or varves—now exposed east of Red Rock Lakes on Red Rock Creek—accumulated at the bottom of the lake.

Pleistocene Lake Centennial's water level fluctuated as it cut through the bedrock sill and slowly dropped, exposing shoreline sand, which was then piled into dunes up to 98 feet high on the northeastern side of the lake by the prevailing winds. Most of the sand accumulated during the late Pleistocene, and it appears that the lake was gone by 12,000 years ago. Incredible fossil finds in the valley show that rodents, pronghorn, deer, bison, bears, beavers, coyotes, wolves, cats, horses, camels, and Columbian mammoths thrived during the last glacial period. The mammoths are particularly well known from preserved remains of their teeth, tusks, and other bones excavated from the Merrell Site near the west end of present-day Lima Reservoir. Radiocarbon dates, obtained by measuring carbon isotopes, indicate that humans were in the valley by at least 10,500 years ago and likely much earlier. Quietly exploring the Red Rock Lakes by nonmotorized boat is an experience of a lifetime, but the lakes are treacherously muddy, so do not get out of your boat.

The geologist is standing on fine-grained lake sediments deposited in Lake Centennial during Pleistocene time. When the lake drained, Red Rock Creek deposited gravel on top of the lake sediments. The geologist's right hand is at the transition. —Courtesy Ken Pierce

The Centennial Mountains, which form the Continental Divide, are reflected in the shallow waters of the Red Rocks Lakes east of Lima. The Centennial fault lies at the base of the mountains. —Courtesy John Lambing

Madison River Terraces

Giant Steps South of Ennis

In the Madison Valley south of Ennis, giant steps climb up from the Madison River. The stair treads are all relatively level and separated by risers, but the landscape carpenter (the river) did not make them a uniform height or width. These terraces, remnants of former valley floors, were left behind as the river eroded down through its floodplains. Some steps are paired, meaning they occur at similar elevations on opposite sides of the river, like those in the central part of the Madison Valley. Where the terraces are unpaired, like those downstream from Quake Lake, the forces of erosion and deposition were not uniform across the river valley, resulting in different outcomes on each side. Most of the terraces are covered in gravel deposited by glacial outwash streams during Pleistocene time.

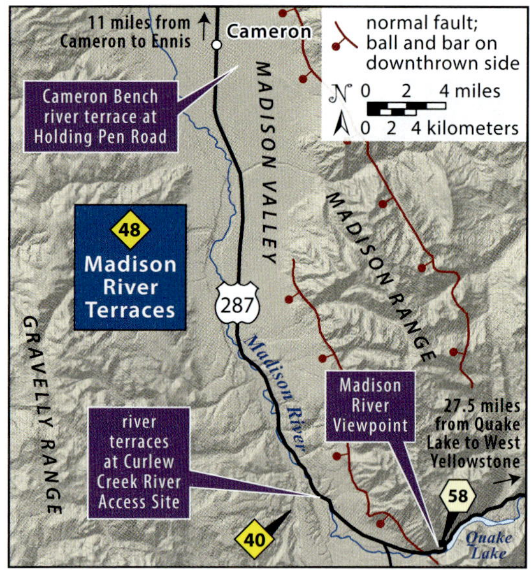

From the Madison River Viewpoint at Earthquake Lake Visitor Center, travel 7.4 miles west on US 287 to the Curlew Creek River Access Site to see multiple river terraces. Continue north another 23.2 miles to just south of Cameron to see large terraces.

The formation of river terraces is complex, but in simplest terms, they form by erosion or sediment accumulation. For example, when a river erodes through bedrock, it can cut terraces and deposit a thin veneer of gravel (alluvium). If a valley already exists, like the big, fault-bounded Madison Valley, it might fill with alluvium until a change in stream power causes the river to erode into the material, leaving a terrace. As the river erodes downward, many levels of terraces can form as the river pauses in its downward progression, migrates back and forth to make a flat floodplain surface, then erodes again. If gravel accumulates faster than the river can transport it, the valley might partially fill, leaving a terrace behind when downward erosion resumes. In each case, the oldest terrace is the highest, and those in the Madison Valley show us that the river has been cutting downward into the valley over time.

Up to twelve terrace surfaces exist in the Madison Valley, with the highest and oldest, the Cameron Bench, located 250 feet above the Madison River. Glaciation in nearby mountains and over the Yellowstone Plateau was essential to the formation of the Madison River terraces. The Cameron Bench is at least 75,000 years old and might be composed of outwash from the Bull Lake glaciation. The lower terraces formed during the Pinedale glaciation. Climate variations during the ice ages probably provided the variations in river flow and sediment supply that led to terrace formation. The river also incised its channel because the lowest point to which the river could erode was lowered. At the head of Bear Trap Canyon (site 3), where hard basement rock that resists erosion is exposed due to active normal faulting, it has caused both sediment ponding and an increased erosional gradient of the valley. To top it off, the Madison River drainage is on the margin of the Yellowstone hot spot, so thermal bulging of the ground around the hot spot increased the gradient and must have played a role in the downward erosion by the river. The abundance of terraces in the Madison Valley may argue for an abnormally rapid rate of uplift compared to other glacial drainages in the region.

Madison Valley stream terraces along the west bank of the Madison River looking west from US 287 a few miles south of Cameron.

Madison Valley stream terraces viewed west from US 287 near the Curlew Creek River Access Site, just a few miles north of Earthquake Lake. At least four terraces are easily identified, with a small floodplain at the level of the Madison River. If visiting the Earthquake Lake Visitor Center, the Madison River Viewpoint is a good place to see recent home development on the terrace surfaces.

49 Lewis and Clark Caverns

Interior Decorations of the Underworld

Lewis and Clark Caverns became Montana's first state park in 1937. The limestone caverns are at least 600 feet deep and boast three-quarters of a mile of accessible passageways you can tour with a guide. The origin of the caverns started with the deposition of Madison Group limestone in a tropical ocean that covered the area from 358 to 325 million years ago. Look for button-shaped crinoid and horn-shaped coral fossils in the rocks along the walking path to the cavern entrance. The rocks were folded and faulted during Laramide compression, forming fractures in smaller folds that would later influence the location of the caverns. A few million years ago, when crustal stretching started to uplift the mountains we see today, the ancestral Jefferson River started to cut down as the land rose, initiating cavern formation.

The caverns formed as rainwater incorporated carbon dioxide from the air and decaying organic matter in the soil to form a weak carbonic acid that percolated down to the water table. The slightly acidic groundwater slowly dissolved openings in the bedding and fractures within the cores of small folds in the limestone. The caverns concentrated where the groundwater perched above impermeable layers of a muddier limestone in the lower Madison Group, preventing cavern formation from occurring any deeper. Eventually, as the river cut deep into the rising land, the groundwater table dropped, and the caverns drained. They are now 1,300 feet above the river. The caverns likely formed in Pleistocene time, and once drained, the roofs collapsed, leaving rubble piles and arched ceilings.

From just north of Cardwell, travel 7.3 miles east on MT 2, and turn left (north) at the Lewis and Clark Caverns Main Visitor Center. Travel 6.3 miles up to the Cave Visitor Center for the tour (reservations required). The tour is physically challenging, beginning with a 0.75-mile-long walk on a trail that climbs 300 feet to the entrance. Once inside, the tour takes 2 hours, and some of the passages are steep and narrow. It is 50°F in the caverns year-round, so bring appropriate clothing.

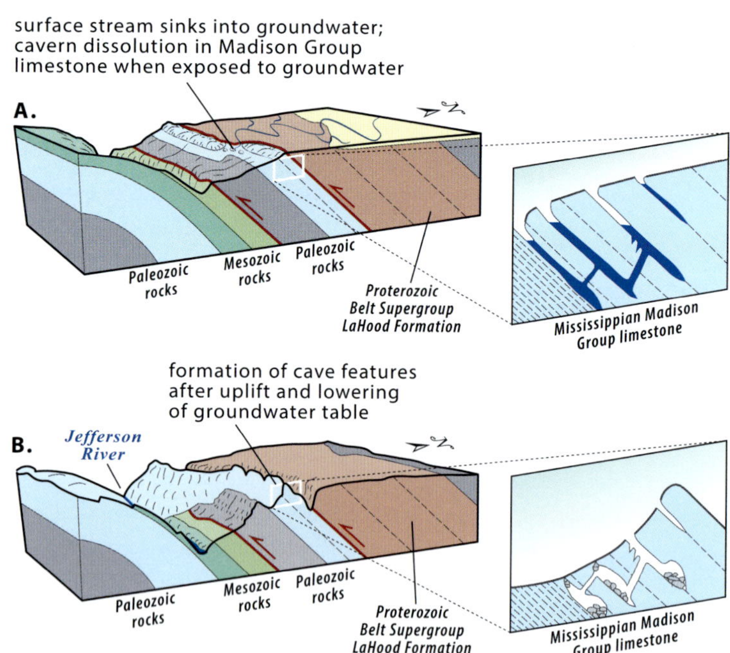

The caverns formed when groundwater dissolved them, and the cave features formed when the Jefferson River Canyon was cut, and the caverns were left high and dry. —Courtesy of Rich Aram

The speleothems, or cave formations, formed when water carrying dissolved limestone seeped inside the drained caverns, released carbon dioxide (making it less acidic), and precipitated crystals of calcite. As the water slowly dripped, it formed icicle-like stalactites that hang from the ceiling, stalagmites that grow upward from the cave floor, and columns where they join. Each drop of water can deposit a layer of calcite, which are visible in broken speleothems. The rate of growth can vary greatly, but typically, it takes thousands of years to accumulate just a few centimeters of calcite.

Flowing water deposited sheets of calcite, making ribbons on the walls called cave bacon due to reddish layers containing organic compounds leached from the overlying soils.

Although the caverns are mostly dry, this shallow pool of turquois-colored water is a highlight of the guided tour.

Cross section of a stalagmite, about 8 inches across, showing the outward growth layers of calcite precipitated from dripping water.

If you ask your tour guide, you just might be offered the opportunity to see the remains of this cave rat entombed forever in flowstone on the cave floor. The rat remains are about 12 inches long.

50 Ringing Rocks

Hardrock Music on a Boulder Pile

An unusual pile of igneous rocks west of Whitehall is known as Ringing Rocks because they ring like bells. The rocks are one of many intrusions that compose the Boulder batholith, a mostly granitic body of rock that formed around 76 million years ago. The Ringing Rocks are not granitic but rather are a hard, fine-grained, brown-weathering, dark igneous rock called gabbro. This small intrusion forms a bullseye-pattern with the dark-colored rocks surrounding a core of lighter-colored granite.

The rock pile is an accumulation of huge, angular rocks that cover the ground as though dropped there by some giant hand. Between the rocks are open spaces. The pile is best described as a *felsenmeer*, a German term meaning "sea of rock." It formed in place by freezing and thawing and weathering along fractures that broke once-solid rock into the large boulder field. At Ringing Rocks, some incompletely weathered fractures transition to completely weathered fractures that separate the stacked rocks, confirming they formed in place. Felsenmeers are typically found in mountain environments like this one, where frost wedging is prevalent. The pile likely formed during the cold climates of the Pleistocene.

This rock features incompletely weathered fractures, showing the pile formed in place. With time, these rocks will likely become completely separated but still touching within the pile.

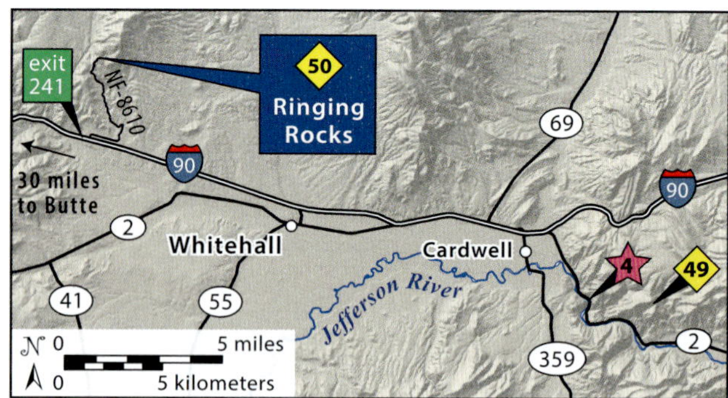

From I-90 at Whitehall, travel 7.1 miles west, and take exit 241 near Pipestone. Take the first right turn (east) on NF-8610. At 1.5 miles, stay left (northwest) on the dirt road for 2.8 miles. Turn right (east), and travel northeast for one-third mile to the Ringing Rocks parking lot. The road is rough, so access the site with a four-wheel-drive, high-clearance vehicle.

Most of the rocks in the pile ring when struck by a hammer, and striking different-sized boulders provides distinctly different tones. It's uncertain what causes the ringing, but it seems to be related to the rock composition, the way the rocks were piled, and the tight, interlocking, fine-grained minerals. If a rock is removed from the pile, it doesn't ring! If you strike a rock that is not in the pile, it will make a thudding sound. The reddish-brown weathering rind shows that the rock has iron in it, suggesting that iron content has something to do with the ringing, but testing of rock piles around the globe of similar composition and iron content show that one pile will ring, while another will not. There is evidence that the formational stress preserved within the rock is important. A limp guitar string will not resonate, but one that is under stress will make sound. Perhaps rocks that retain relict stress from their formation deep below the surface are more apt to ring. Studies show that ringing rocks expand when removed from the pile, while those that thud do not.

Park at the informational sign, and walk the short distance up the road to the site. A rack there contains hammers for the public to use. The rocks break, so tap them lightly. The best spots to make the rocks ring like a bell are obvious by dings on the rocks from repeated hammer strikes. Please do not move or break the rocks because such a disruption might well bring an end to the music.

The hard, fine-grained gabbro has a distinctive, brown-colored weathering rind.

At Ringing Rocks west of Whitehall is a pile of rocks that ring like a bell when struck with a hammer. In the distance at left are more typical granitic rocks of the Boulder batholith.

51 Great Falls of the Missouri River

Postglacial Dissection of the Plains

The Great Falls of the Missouri River are named for five waterfalls that drop the river more than 612 feet over 10 river miles downstream of the town of Great Falls. The falls located farthest downstream, called Big Falls (Great Falls), formed a cascade 87 feet high and 900 feet wide at peak flow. The cataracts were well known to the Mandan, who named them *Minni-Soze-Tanka-Kun-Ya* or "the great falls." The Mandan told Lewis and Clark about the falls, and the expedition encountered them in the summer of 1805, with Lewis noting the "agreeable sound of a fall of water" and "the spray arise above the plain like a column of smoke." It took the expedition thirty-one days to portage around the falls, requiring them to build wagons and haul canoes and supplies.

The Great Falls did not exist before the glacial advances of the Pleistocene, when the Missouri River flowed northeast into Hudson Bay. The preglacial river flowed south and east of the current falls along Sand Coulee Creek before turning north to Belt Creek. When the Laurentide ice sheet spread out of Canada, the river could no longer flow north and followed roughly the southern margin of the ice. When the ice started to retreat around 13,000 years ago, the river finally began down the path that would create the Great Falls of the Missouri.

The river initially cut down through soft glacial deposits, but it got stuck in the underlying hard layers of flat-lying sandstone and shale of the Cretaceous Kootenai Formation, forming falls over the more resistant ledges of sandstone. Big Falls (or Great Falls) formed over a particularly thick, hard layer of marine sandstone deposited about 120 million years ago in the Western Interior Seaway. The other falls, all

The Great Falls in the summer of 1880, long before Ryan Dam was built here in 1915. —Courtesy of Haynes Foundation Collection, Montana Historical Society

The Great Falls as they look today, tamed by Ryan Dam for hydroelectricity.

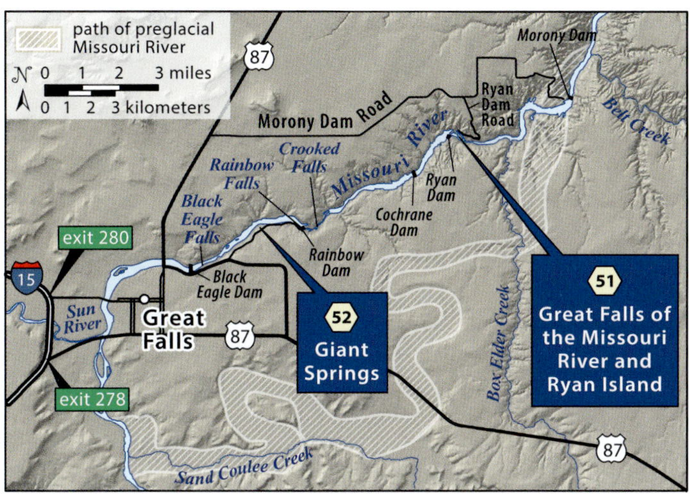

From Great Falls, travel 3.7 miles north on US 87, and turn right (east) for 6.8 miles on Morony Dam Road. Turn right (south), and travel 1.8 miles on Ryan Dam Road to its terminus below the dam. In season, you can cross the bridge onto Ryan Island, which has several geological interpretive signs.

upstream of Big Falls, cascade over thinner layers of sandstone interbedded with mudstone deposited by streams on the Cretaceous coastal plain. Over time, the falls migrated upstream as falling water eroded the soft rocks underneath the sandstone ledges. Each time the fractured, unsupported sandstone collapsed, the lip of the falls moved upstream. The headward erosion has stopped now that four of the five falls are dammed for hydroelectric power.

At Ryan Dam, the river is more than 340 feet below the surface of the plains. Some of the erosion might be from glacial meltwater during deglaciation, but rivers without glacial inputs are also eroding into the plains. Rivers can erode downward when their lowest point drops, in this case, the Gulf of Mexico, but the oceans have been rising in the postglacial period. The most likely explanation is that the plains are rising, increasing the river gradient, probably because the plains are on the eastern flank of rising land in the Basin and Range province. As the brittle crust stretches and thins, the ground rises in response to rising heat and reduced load on the underlying mantle.

The Missouri River, here viewed downstream from Ryan Dam, has been dissecting through hard sandstone and soft shale of the Cretaceous Kootenai Formation since the Pleistocene ice retreated 13,000 years ago.

52 Giant Springs

Good Work of the Artesians

See map on page 123.

Giant Springs is the centerpiece of a little state park on the south bank of the Missouri River near the eastern edge of the town of Great Falls. Although estimates vary greatly, perhaps around 200 million gallons of water per day well up through fractures in the Cretaceous Kootenai Formation sandstones and pour onto the surface at the springs. The spring water then flows a short 201 feet down the Roe River and into the Missouri River. The Roe River was officially recognized as the world's shortest river by the Guinness World Records in 1989 and held that title until 2000, when they eliminated the category following a dispute over the length of the D River in Lincoln City, Oregon, whose length fluctuates with the tides. What is not short is the distance the water traveled underground through limestone of the Madison Group to get to the spring, at least 40 miles north from the Little Belt Mountains.

The age of the groundwater that emerges at Giant Springs was once thought to be about 3,000 years old, but that determination is now in question due to a reevaluation of the dating technique. However, water emerging from the Madison Group elsewhere in Montana has been shown to be thousands of years old using isotopes like chlorine-36, which record how long the water has been underground and isolated from the atmosphere. This fossil water is a gift from the past and can be depleted because it takes so long for nature to put it back in the rocks.

The water traveled from the Little Belt Mountains by following the Madison Aquifer, a thick body of porous and permeable limestone in the Mississippian Madison Group. This unit is loaded with dissolution features ranging from small holes to large caverns that now hold huge quantities of groundwater. Over the millennia, rainfall and snowmelt seeped down into the rock, filling the limestone caverns with groundwater. Sandwiched between impermeable layers of rock, the groundwater flows downslope in the limestone more than 4,000 feet in elevation to the Missouri River, where it intersects fractures in the overlying, younger rock and gets pushed 700 feet to the surface by the hydrostatic pressure of the water. The emerging water is called an artesian spring because it comes out of the ground under pressure and requires no pumping to get it to the surface.

Water emerges from several fractures in the area, resulting in many springs, including some that come up into the bed of the Missouri River. The Giant Springs has a constant temperature of 54°F and so served as a source of water for the First People during cold spells when the river and other surface water was frozen. In 1805, Meriwether Lewis wrote, "I think this fountain the largest I ever beheld, extremely transparent and cold, very pure and pleasant." The water is still relatively pure and has been used as a source of bottled water. Look for plant fossils preserved as black impressions on the sandstone bedding surfaces of the Kootenai Formation around Giant Springs.

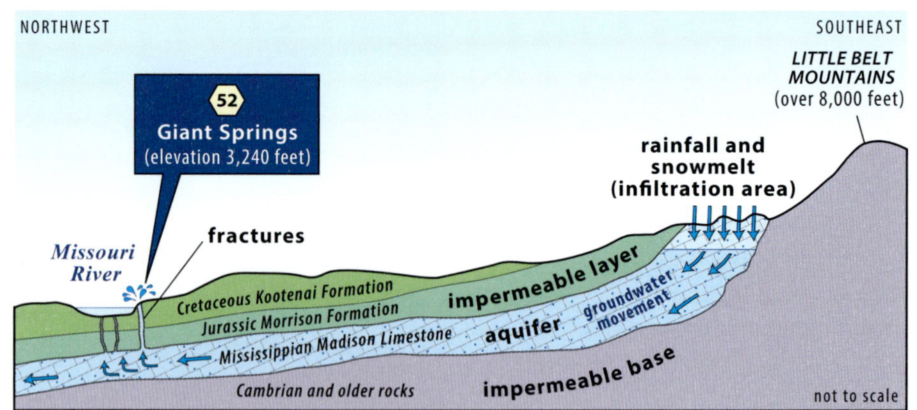

Sketch showing how water traveling downslope under pressure is pushed to the surface through fractures to form this artesian spring. —Modified from Giant Springs Bottled Water Company

The water emanating from Giant Springs along the banks of the Missouri River (at left) travels from the Little Belt Mountains following an aquifer in limestone of the Mississippian Madison Group.

53 First Peoples Buffalo Jump

An Early Use of Montana Geology

For millennia, the First People hunted buffalo by driving them over rocky cliffs. It is estimated that as many as thirteen million buffalo once lived in Montana, so it is no surprise that more than three hundred known buffalo jumps are scattered across the state. Only three sites are protected: the Madison Buffalo Jump near Three Forks, Wahkpa Chu'gn near Havre, and First Peoples Buffalo Jump State Park north of Ulm, Montana. The latter is one of the largest buffalo jumps in North America based on its areal extent. The high, grassy meadow is bordered on three sides by a horseshoe shaped sandstone cliff that ranges from 30 to 50 feet high. As many as fourteen different tribes used the jump, probably due to the optimal geology and its proximity to the Missouri River.

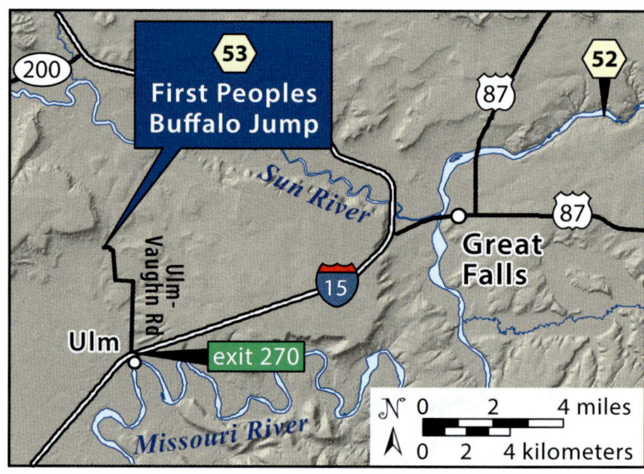

From Ulm, travel 3.7 miles north on Ulm-Vaughn Road. Turn left (west) at the sign for First Peoples Buffalo Jump State Park. Walk the 3-mile Loop Trail to the cliff rim and up to the prairie dog town atop the bench.

The rock that forms the cliff walls of the buffalo jump is the Taft Hill Member of the Blackleaf Formation, a cross-bedded, quartz-rich sandstone with scattered, green-colored grains of glauconite, a clay mineral. The sand was deposited in coastal beaches and shallow seas during Cretaceous time, around 110 million years ago. In the cliff wall, look for low-angle cross-bedding formed by migrating and shifting sand, broad sandy hummocks piled up by storm waves, dark mud chips ripped up during storms, and curved ripples formed by gentle currents. The rocks atop the buffalo jump are covered with these stone ripples.

Erosion is the primary reason this horizontal layer of sandstone served as an excellent place to run buffalo off a cliff. The horizontal sandstone layers and soft shale above and below were once continuous across the plains and mostly undisturbed by tectonic forces. Following the glacial period, as the plains began to rise and the Missouri River began to cut downward, these sedimentary rocks were eroded as the river migrated back and forth. Had all the rocks been soft shale, there would be no buffalo jump here because the river would have leveled the landscape. The well-cemented sandstone resisted total erosion, forming a step in the landscape.

A depth of about 18 feet of compacted bison bones lies at the base of the cliff, demonstrating its long use. Prior to the introduction of horses by the Spanish, buffalo jumps were essential to killing bison, which are fast and weigh between 1,000 and 2,000 pounds each. The most widely accepted scenario is that hunters would encircle a bison herd miles away from the jump and drive them toward the cliff. Stone cairns and fences on the high meadow were used as drive lines to guide the animals. At some point, a fast runner, perhaps disguised by wearing an animal hide, would race ahead of the herd to help lead them toward the cliff. The runner would leap over the edge where there was a safe place to shelter as the buffalo plummeted over and into the jagged rock debris below the cliff. Any animals that did not perish during the fall were likely maimed and easily killed. This ancient place deserves our quiet contemplation and respect.

Ripple marks on bedding surfaces of the sandstone were formed by gentle currents at the edge of a sea in Cretaceous time.

Hard, horizontal layers of sandstone form the First Peoples Buffalo Jump north of Ulm.

54 Alder Gulch

A River of Gold

On May 26, 1863, a group of prospectors traveling back to the gold fields of Bannack (site 20) discovered gold in Alder Creek. They had hoped to pan enough gold to buy tobacco, but they found a bit more. Within three months, so many mining camps sprung up in Alder Gulch that it was referred to as the Fourteen Mile City. The largest of the camps was known as Verona. On June 16, 1863, a claim was filed on 320 acres to be used as a townsite named Varina, after Varina Davis, the First Lady of the Confederate States of America. When the charter was presented, a Unionist judge objected and changed the name to Virginia. By 1864, Alder Gulch drew a flood of miners, peaking at more than ten thousand. By February 7, 1865, Virginia City became the capital of Montana Territory and boasted a population of more than five thousand people.

In the beginning, miners used hand tools and sluice boxes to separate the placer gold from the stream gravels. In a sluice box, heavier grains of gold catch on horizontal slats as water flushes sand and gravels down the length of the box. In 1867, hydraulic mining—in which pressurized water washes gravel from banks and carries it through sluices—was employed as claims were consolidated, and more money could be spent to remove the gold. By 1889, an estimated $90 million in gold had been extracted, a figure equivalent to no less than $40 billion today. In 1898, new life was breathed into the placer mining with the arrival of the Maggie A. Gibson, a steam-powered dredge barge that was able to dig deeper and extract gold that hydraulic mining methods could not reach. The dredge barge made its own pond by digging the gravel in front and dumping the waste rock behind. A sluice box on the boat removed the gold, and a conveyer belt swung back and forth, dumping the dredge spoils into rows in the creek's channel.

Conrey electric dredge on Alder Gulch in 1911. The tailings boom on the right spewed out wasted gravel, making piles as it moved back and forth. —Photo by Hennen Jennings, US Bureau of Mines

By 1906, electrically powered dredge barges joined the fleet, including Dredge No. 4. Built in 1911 at a cost of $296,000, it was reported to be the largest dredge in the world. By 1922, the entire floodplain of Alder Creek had been consumed, resulting in an additional $10 million in gold. Virginia City was never dug up because it was not built on gold-bearing gravels, but the original town of Nevada City was removed by dredging in the early 1900s. Many of the old cabins you see today in Nevada City

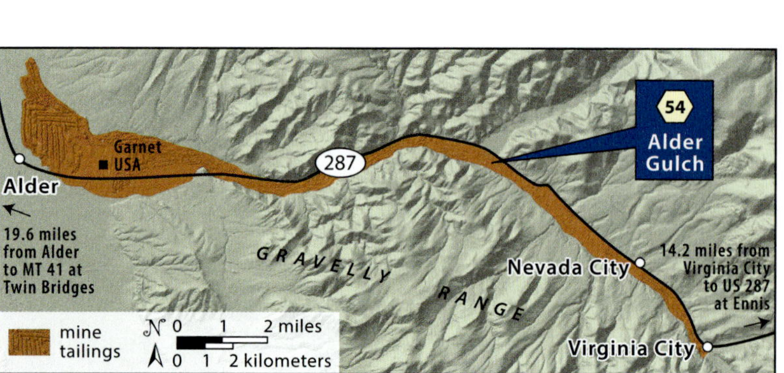

Take in the many rows of dredge spoils or mined gravels mostly south of US 287 in the 8 miles between Virginia City and Garnet USA.

were moved to their present position from elsewhere in Montana by Charles and Sue Bovey. Charles was a legislator and rancher who took a keen interest in preserving this historic treasure. Although some remediation is ongoing in Alder Gulch, many furrowed rows of gravel and dredge ponds remain along MT 287 west of Virginia City.

From the start of the gold rush, lode or bedrock mining took place in the upper reaches of Alder Gulch, but it was never as productive as the placer mining. The gold in the gulch's gravels must have come from sources in the hard rock, but the exact location and the geologic processes that produced it are not known. The limited gold in the hard-rock mining district occurs in quartz veins that intruded Archean metamorphic rocks, including some loaded with garnets. The source and age of the vein mineralization is uncertain, but it's likely related to the intrusion of the Tobacco Root batholith in Late Cretaceous time. Today, few people live in Alder Gulch, but mining continues for the hard almandine garnets, known as red gold, that were once discarded as a nuisance to the gold miners. The garnets are now mined from tailings and bedrock, primarily for use as industrial abrasives.

Tailings left from gold mining on Alder Creek near Virginia City over a century ago. The ridges of boulders are the waste products dumped by a floating dredge boat as its boom waved back and forth.

55 Butte
The Richest Hill on Earth

Butte started as a gold camp in 1864, but gold was scarce, and the silver deposits were awash in copper, which had little value at the time. That all changed in the late 1800s with the rise of big electrical industries hungry for copper wire. By 1898, Butte was producing much of the world's copper. The ore was concentrated in steep veins, some rich in silver and manganese, others in copper, and both requiring underground mining. The deepest mine reached 5,291 feet, making 5,549-foot-high Butte a mile high and a mile deep. One mine was 135°F, and school kids delighted in sweaty miners going up in smoke as they emerged into the cold, surface air. The rich veins were worked out by the 1950s, but Butte pressed on with open-pit mining of lower-grade ore that produced more waste than ore. By the turn of the twenty-first century, 211 billion pounds of copper and 3.1 billion pounds of manganese had been produced from what was known as the richest hill on Earth.

The ore minerals were deposited between 66 and 62 million years ago into 76-million-year-old granitic rocks of the largest of several intrusions that make up the Boulder batholith (site 22). Hot, circulating water deposited the ore minerals into cracks (veins) and breccia zones in the rock. As fluids moved outward from the cooling magma, minerals were deposited in zones, with copper-rich ores in the center and lead-zinc-silver ores in the surrounding rocks. Long after the intrusion cooled and the ore veins formed, warm, near-surface water leached minerals and precipitated low-temperature ores, like chalcocite, a black copper sulfide, in an enrichment zone just below the water table.

There are more than 10,000 miles of abandoned tunnels below Butte, and the old Berkeley Pit is the deepest lake in Montana at more than 900 feet. When underground mining stopped, the pumps that kept the tunnels dry were turned off, and the pit began to fill with groundwater. Abundant sulfide minerals created acidic water that leached metals like copper, arsenic, cadmium, zinc, and lead, creating a toxic soup of acidic water and metals in the pit lake. The pit water is so toxic that it poisoned more than 3,000 snow geese that landed on it in 2016. Perched above the Berkeley Pit is the Yankee Doodle Tailings Pond, a 2.5-square-mile wet tailings pile with an earthen dam about 650 feet tall. The ore was once smelted in Butte, but most operations were moved to the big smelter in Anaconda in the 1890s. Pit mining continues to this day for copper and molybdenum brought closer to the surface over time by the active Continental normal fault. The waste rock generated raises the Yankee Doodle Tailings Pond by about 5 feet each year.

Most of us have been touched in some way by Butte copper, and so perhaps the cleanup of the mining contamination is a societal responsibility. Federal Superfund money has been invested to remediate the Berkeley Pit, the Anaconda smelter, metals along Silver Bow Creek, the floodplain of the upper Clark Fork River, and sediments that collected behind Milltown Dam near Missoula. Water in the Berkeley Pit is pumped to prevent it from contaminating the groundwater under the city of Butte, and it is treated and discharged into Silver Bow Creek. Lime is added to the slurry of tailings pumped from the active Continental Pit to reduce its acidity, which has also decreased the acidity of the Berkeley Pit water. Mine-waste dust is another long-term problem, so efforts to cap and vegetate the tailings in Butte are an ongoing containment process.

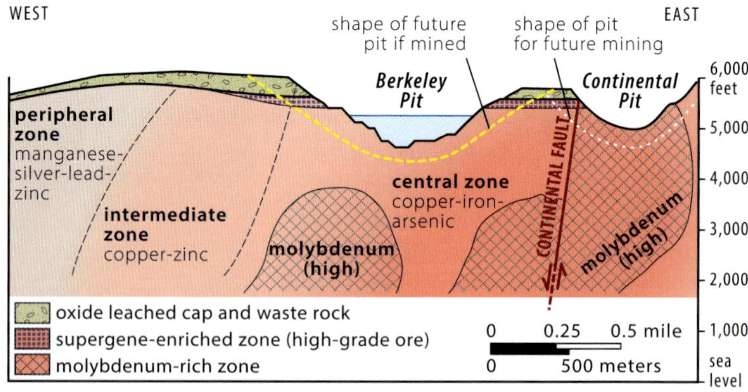

Simplified cross section through the Butte mining district showing the different zones of mineralization. —Modified from Gammons and other, 2006

Metal-contaminated water in an animal print on the floodplain of the Clark Fork River in the Deer Lodge Valley. Flooding in 1908 deposited huge quantities of metals from Butte, leaving behind dead zones called "slickens" on the floodplain.

From I-90 in Butte, travel 1.2 miles north on Montana Street. Turn right (east) on East Mercury Street for about 1 mile. Turn right (southeast) on Shields Avenue, and in 0.2 miles turn left (north) into the parking lot for the Berkeley Pit Viewing Stand. Consider stopping at the Butte Overlook (45.9911, −112.4751) on I-15 southbound from Helena. Aerial view of the Berkeley Pit, Yankee Doodle Tailings Pond, and the Continental Pit in 2025.
—ESRI image courtesy of John Sanford, Montana Bureau of Mines and Geology

N
0 0.5 1 mile
0 0.5 1 kilometer
Yankee Doodle Tailings Pond
Berkeley Pit
Continental Pit
55
Butte (Berkeley Pit Viewing Stand)
15
East Mercury Street
Shields Ave
Montana Street
Butte
exit 126
90
Butte Overlook (southbound lane access only)
0.8 miles from Butte Overlook to I-90 interchange

56 Libby Vermiculite Mine

A Serpent in Paradise

Like most tragic stories, the sad ending in Libby could not have been foreseen in the beginning. In the 1860s, gold prospectors found a bit of color in Libby Creek, bringing miners and railroads to the region. Many years later, local rancher and prospector Ed Alley found an odd mineral while exploring a mine tunnel north of Libby. He noticed that the heat of his candle caused the material to swell up and turn gold in color. He had found the mineral vermiculite, and so in 1919, he bought the Rainy Creek claims and started the Zonolite Company to market his product for use in insulation, plaster, and gardening soil. Starting in 1925, vermiculite was mined from a big, open pit high on Vermiculite Mountain, changing the town and the people of Libby forever.

Vermiculite is a platy clay mineral commonly formed by the alteration of micas like biotite. It expands when heated because trapped water becomes steam that puffs it up like popcorn to about fifteen times its original volume. Not all vermiculite is dangerous, but the vermiculite mined in Libby contained asbestos minerals. The vermiculite was mined from the Rainy Creek intrusion, a pyroxenite-syenite igneous rock

Scanning electron microscope image of elongate amphibole minerals, some of which are asbestiform, collected from attic insulation from Libby, Montana. —Courtesy of the US Geological Survey

From Libby, travel 5.5 miles northeast on MT 37 to the Rainy Creek Road. The mine is blocked off by a large gate 0.8 miles up the road.

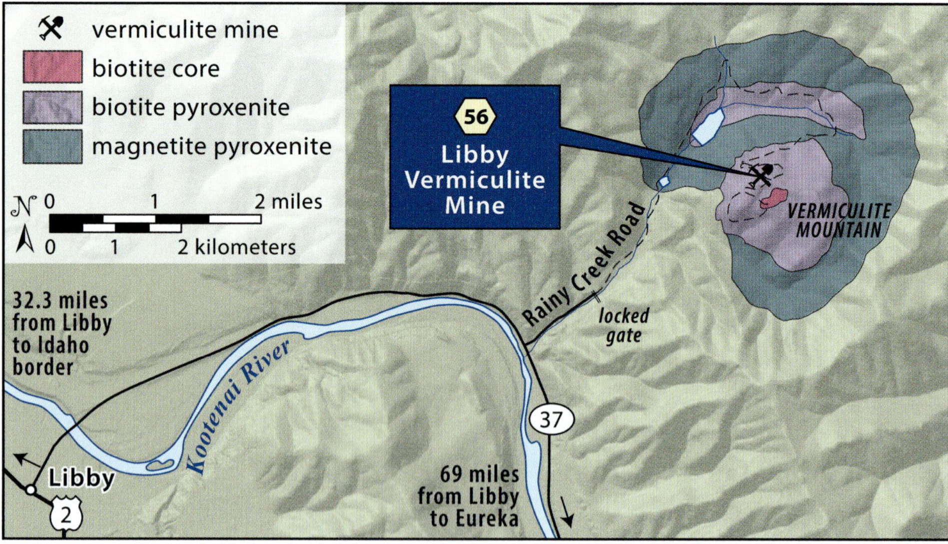

intruded into the Mesoproterozoic Belt Supergroup rocks, perhaps late in Cretaceous time. Pyroxenite is rich in iron and magnesium minerals typically derived from partial melting of the mantle. The intrusion has a small biotite core surrounded by biotite pyroxenite and magnetite pyroxenite. The biotite pyroxenite rocks were altered to vermiculite by groundwater or hydrothermal fluids, superheated water containing dissolved minerals. Dikes of white syenite, an igneous rock composed mostly of feldspar, appear to have altered the pyroxenite to tremolite, winchite, and richterite, dangerous types of hard, long, thin amphibole group crystals called asbestos.

By 1963, when W. R. Grace bought the company, the mine was producing more than 80 percent of the world's supply of vermiculite pellets being sold as Zonolite insulation for use in homes. Libby's claim to fame turned to infamy when miners, their family members, and those in the community began falling prey to dreadful lung diseases from asbestos exposure. The microscopic asbestos needles were released into the air as dust when the rock was processed and crushed. The needles then lodged in people's lungs, causing mesothelioma, lung cancer, and asbestosis, a chronic and deadly lung disease. Miners carried it home on their clothes, and it was spread intentionally on playgrounds and ballfields to soak up water after it rained. It was everywhere.

The mine shuttered in 1990, but the damage to human health had already been done. As of 2021, at least seven hundred residents of Libby had died from exposure to asbestos, thousands more are severely ill, and many more will die as a result. Evidence presented to a grand jury showed that W. R. Grace was aware of the health risks posed by the vermiculite as early as the late 1970s and yet did not act on the data. In 2009, a criminal trial followed, but the company and three executives were acquitted of charges that they knowingly exposed workers to asbestos. In 2002, the Libby Asbestos Superfund Site was established to remediate thousands of properties, remove millions of cubic yards of contaminated soil, and treat the soil and water.

Processed vermiculite pellets packaged and sold as Zonolite Insulation for use in homes across America. Scale bar is in centimeters.

At the height of operations (1963 to early 1980s), the Libby Mine was spewing up to 5 tons of asbestos-laden dust into the air above Libby each day. —Public domain photo courtesy McGarvey Law

57 Fort Peck Dam

Power to the People

In 1933, many folks were on hard times and looking for work due to the Great Depression. When the federal Public Works Administration approved a dam construction project on the Missouri River in Montana, the hopeful flocked to the region. This remote part of eastern Montana had few services, so the town of Fort Peck was built for the Army Corps of Engineers, and shanty towns like Wheeler, Delano, McCone City, and a dozen others sprang up to house the workers. At its peak in 1936, the New Deal project provided jobs for 10,546 people. What they built was a dam 21,026 feet long and 250 feet high with a present generating capacity of 185 megawatts of electricity, enough to power 147,000 homes. At 134 miles long, Fort Peck Lake is the fifth largest reservoir in the country, with more shoreline than California has coastline.

The massive sediment-and-rock-fill dam has a clay core between two permeable shells of coarser material. Gravel toes were built at the base of each shell to anchor each side of the dam. There was not enough hard rock in the area to build the toes, so shonkinite was brought in by rail from Snake Butte, a laccolithic intrusion 115 miles to the west, southwest of Harlem. Shonkinite is a dense, tough, basalt-like rock with tightly interlocking grains, just the stable characteristics engineers were looking for in fill material. The dam was situated on 160 feet of river deposits resting on 1,000 feet of Bearpaw Shale. To prevent water from getting under the dam, slabs of sheet steel were driven more than 150 feet into the Bearpaw Shale to create a waterproof barrier the length of the dam. The fill for the dam was hydraulically dredged upstream and sent to the construction site as a slurry through pipelines.

Building the dam atop the Bearpaw Shale was a dubious choice because the old Western Interior Seaway mud is notoriously unstable. The organic, clay-rich, black shale is very platy so has built-in planes of weakness not suitable for a dam foundation. To top it off, the oceanic mud accumulated airfall ash from western Montana volcanoes that weathered to layers of bentonite, an infamously unstable, swelling clay that can increase in volume by a factor of eight when wet. It should be no surprise that on September 22, 1938, with the dam partway finished, cracks appeared on the upstream side of it, and the east end failed, allowing 5 million cubic yards of material to spill into the growing lake. The landslide moved at about 10 miles per hour, and thirty-four workers were carried with the sliding material. Eight died, and six bodies remain permanently entombed at the base of the dam. Analysis of the failure showed that the weight of the dam material probably caused slippage on the expansive clay seams in the underlying Bearpaw Shale. At the time, a board of consulting engineers conceded there was risk to completing the dam, yet they still recommended finishing the project.

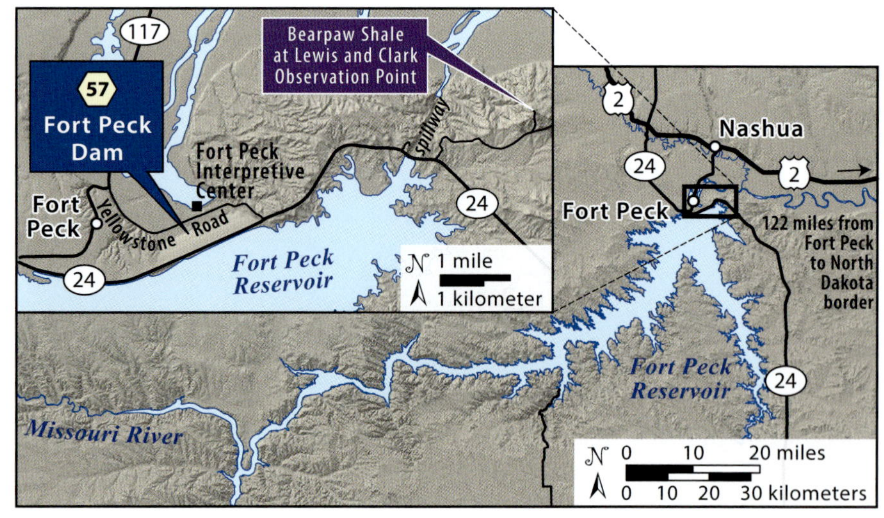

MT 24 runs along the crest of the Fort Peck Dam, and overlooks and interpretative signs provide information. The Fort Peck Interpretive Center lies at the base of the dam along Yellowstone Road. To see outcrops of the Bearpaw Shale, head east on MT 24 from the spillway, and take the first left (north) on unmarked Fort Peck Tower Hill Road for 1.9 miles to the Lewis and Clark Observation Point.

Fort Peck Dam spillway, a classic piece of Art Deco architecture, under construction. —Courtesy Robert Etzel, US Army Corps of Engineers

View to the northeast of the landslide on the east side of Fort Peck Dam that took the lives of eight men on September 22, 1938. —Courtesy Robert A. Midthun, US Army Corps of Engineers

Completed in 1940, the dam has held for eighty-five years, likely due in no small part to the shonkinite skirts flanking the dam. Although built primarily for flood control, it provides water for irrigation, recreation, and power generation. It is the largest earth-filled hydroelectric facility in the United States, with five turbines supplying energy to both the western and eastern grids. The renewable energy is delivered to customers via about 7,400 miles of federal transmission lines and ninety substations. The dam is a classic piece of Art Deco architecture made famous by a photo of the spillway taken by Margaret Bourke-White that appeared on the first cover of *Life* magazine in 1936. Today, the dam is owned and operated by the US Army Corps of Engineers, so it is still providing federal jobs, as it did during the hard times of the Great Depression.

58 Quake Lake

A Whole Lotta Shakin' Goin' On

Montana is earthquake country, and one of the largest earthquakes ever recorded in the Rockies struck near Hebgen Lake at 11:37 p.m. on August 17, 1959. The earthquake measured a whopping 7.3 on the Richter scale and triggered a landslide of 50 million cubic yards of debris that buried Rock Creek Campground to a depth of 225 feet, blocking the Madison River. The water backed up and created Earthquake Lake, unofficially known as Quake Lake. Many aftershocks of 5.0 and 6.0 magnitude followed, and in the end, twenty-eight people perished.

Memorialized in the book *The Night the Mountain Fell*, the massive landslide was a disaster waiting to happen. The slide occurred on a steep slope of Archean-age schist, a layered metamorphic rock composed of platy mica grains dipping steeply downslope toward the river. The slope was severely undercut by the Madison River, and so the only thing holding the layered rock on that slope was a thin buttress of dolomitic marble. The buttress shattered during the quake and unleashed the rockslide. In less than 30 seconds, a gigantic block of rock fell into the canyon and slid at more than 100 miles per hour on a cushion of air across the canyon and 400 feet up the opposite canyon wall, burying nineteen people in the campground.

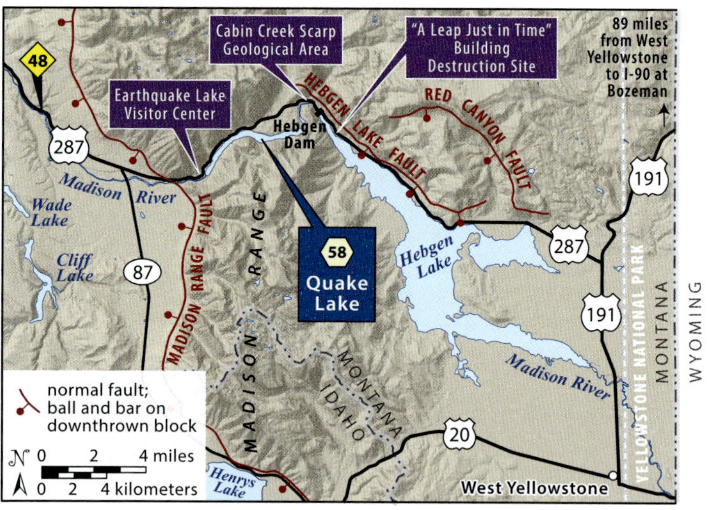

From West Yellowstone, travel 8.3 miles north on US 191. Turn left (west) on US 287, and travel 12.7 miles to the building destruction site ("A Leap Just in Time") on Hebgen Lake. Walk the short trail southeast to the buildings in the lake. Continue US 287 for 1.2 miles, and turn right (north) into the Cabin Creek Scarp Geologic Area. Park near the geological interpretive sign and walk the short trail to the tilted tree and fault scarp. Continue for 5.5 miles along the north shore of Earthquake Lake to the Earthquake Lake Visitor Center.

Three faults—the Madison Range, Red Canyon, and Hebgen Lake—ruptured the ground surface, resulting in fault scarps, steep slopes produced by slip on the fault surface. The three faults are normal faults that raise mountains and drop valleys to accommodate crustal stretching. The Hebgen Lake fault is well exposed as a 21-foot scarp at Cabin Creek about 3 miles east of the visitor center. A trench dug across the scarp by the US Geological Survey showed that the fault had ruptured three times over the last 10,000 to 14,000 years, with the last one producing 10 feet of offset between 1,000 and 3,000 years ago. Some of the scarp we see today may be inherited from that older event. A curious feature at the Cabin Creek fault scarp is a Douglas-fir tree that was tilted during the earthquake toward the fault scarp on the down-dropped side of the fault. It preferred growing straight up, so the change to vertical growth about halfway up the tree happened after 1959.

Hebgen Dam, about 1 mile east of Cabin Creek, dropped 10 feet during the earthquake, cracking the concrete core and spillway, but the dam held. The entire Hebgen Lake basin instantly tilted about 20 feet down to the northeast toward the Hebgen Lake and Red Canyon faults, drowning the northeast shore of the lake and leaving the southwest shore high and dry. Sudden movement of lake water produced a standing wave, called a seiche, that swashed back and forth, topping the dam three times with up to 4 feet of water. Partially submerged buildings can still be observed at the Building Destruction Site on Hebgen Lake about 1 mile southeast of the dam. A truly tragic story unfolded for a

The scar and debris of the Earthquake Lake landslide are clearly visible viewed southwest from a pullout on US 287 a few miles east of the Earthquake Lake Visitor Center. The snags or standing dead trees in the lake died as the water rose behind the slide-debris dam.

family sleeping at Cliff Lake Campground west of Quake Lake. A gigantic block of Huckleberry Ridge Tuff, the deposit from the 2.1-million-year-old Yellowstone eruption, bounded over their picnic table and landed squarely on the parents, leaving their three boys sleeping in a tent nearby unharmed. Another large magnitude earthquake will happen again in western Montana; it's just a matter of when and where.

The Douglas-fir tree behind the student at the Cabin Creek Scarp Geologic Area east of the Earthquake Lake Visitor Center was tilted toward the 21-foot Hebgen Lake fault scarp on the down-dropped side of the fault during the earthquake in 1959. Following the event, it grew vertically.

Historic photo of the Red Canyon fault, which ruptured the ground during the 1959 Hebgen Lake Earthquake, producing up to 20 feet of offset. —Courtesy of the US Geological Survey

59 Chico Hot Springs

Soaking up Earth's Internal Heat

In addition to causing earthquakes, active faults also provide avenues for hot water to rise to the surface and emerge as hot springs. The eighty or so natural hot springs under the Big Sky range from hidden springs in wilderness areas to rustic joints with sketchy pools and destination resorts with five-star restaurants and spas, like Chico Hot Springs in the Paradise Valley. While Yellowstone National Park in Wyoming has a large concentration of hot springs from the underlying magma, most hot springs in Montana, including Chico, are the result of thinned and faulted crust.

Natural hot springs need three things to form: a source of water, a heat source, and a plumbing system that allows water to circulate deeply enough to be heated and brought back to the surface quickly enough to stay hot. The water comes from rain, snow, and groundwater. The heat comes from the geothermal gradient, how rocks get hotter the deeper they are below Earth's surface. For areas far from tectonic boundaries, like Iowa, the gradient is about 80°F per mile. In southwest Montana, the geothermal gradient runs well over 100°F per mile because the crust is extending and thinning, bringing the hot rocks of the mantle closer to the surface. In addition, the closer you get to Yellowstone, the greater the geothermal gradient.

At Chico Hot Springs, the water is heated when moisture falling on the Absaroka Range seeps down along faults and fractures in the rocks to an estimated depth of more than 1 mile, where it gets heated to about 131°F. As it rises under pressure back to the surface, it loses some of its heat to

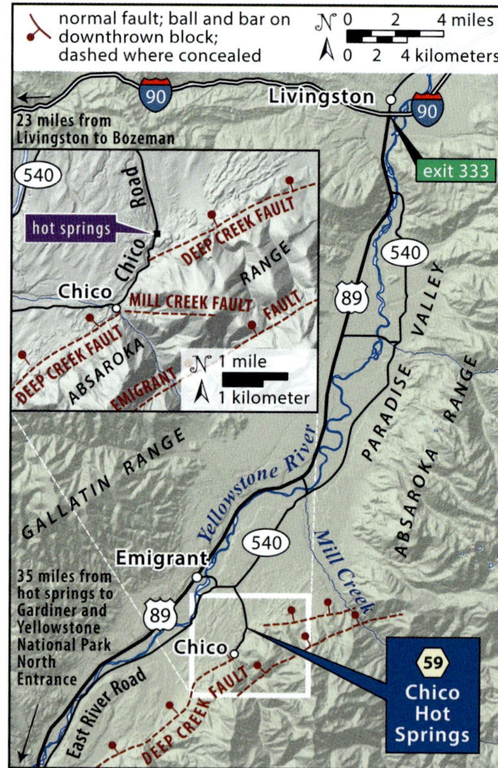

From Livingston, travel 17.4 miles on US 89. Turn left (south) on the Mill Creek Road for 0.8 miles, and turn right (southwest) for 4.9 miles on East River Road. Turn left (south), and travel 1.7 miles on Chico Road to Chico Hot Springs.

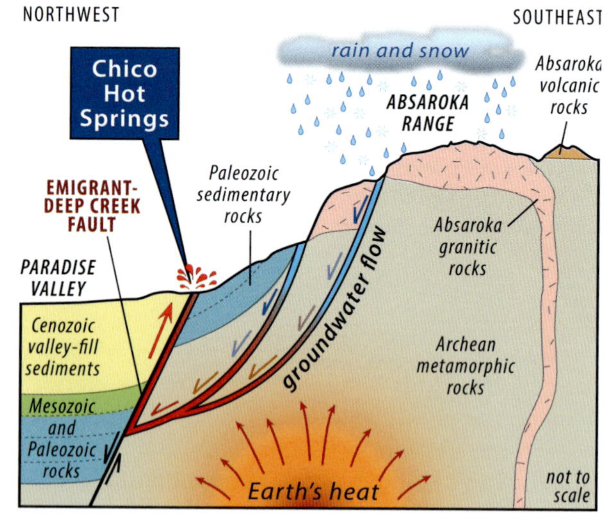

Generalized cross section of the Chico Hot Springs system.

the cooler rocks and near-surface water, emerging from the spring at 113°F. The water gets to the surface by following porous and permeable rock shattered by movement on the Emigrant and Deep Creek normal faults that parallel the mountain front. The water emerges where the rocks are really busted up at the intersection of these faults with branches of the Mill Creek fault. The big normal faults have been raising the Absaroka Range and dropping the Paradise Valley for millions of years, resulting in more than 20,000 feet of offset. We know these faults are still active because the faults have cut the ground surface of the young alluvial deposits along the range front, creating fault scarps.

The hot springs arise from fractures in the Mississippian Madison Group limestone (Mission Canyon Formation), but chemical analyses show that the groundwater probably circulates much deeper through Archean crystalline rocks and only briefly passes through the limestone. Perhaps the northwest-trending Mill Creek normal fault, which cuts crystalline rocks far to the east into the Absaroka Range, conduct the deep hot water westward from the mountains and toward the range frontal faults. The water at Chico Hot Springs has not yet been dated, but water emerging from other hot springs around southwest Montana can be thousands of years old, so it takes time to get from source to surface due to the great depths and distances the water must travel.

The hot spring water, which emerges along intersecting faults at a temperature of 113°F, is allowed to cool slightly before it is piped into the Chico pool.

View east near Emigrant at the Paradise Valley and Absaroka Range. Active normal faults at the base of the mountains provide the conduit for hot water to rise to the surface at Chico Hot Springs.

60 Billings Rimrocks

Hard Times Beneath a Sandstone Cliff

While listening to the turbulent jazz of John Coltrane's *First Meditations*, Billings resident and noted artist Jon Lodge recalled the "roar of a horrendous sound" and "the force of air" as his home was hit by a 10-million-pound boulder that fell from the Rimrocks in Billings in 2010. He survived, as did Coltrane on the CD player, but living below the ever-present danger of a rockfall from the sandstone cliff must be stressful. Fallen blocks and property damage are frequent events below the Rims, with a half dozen damaging rockfalls in the last twenty years. A short walk on the Rimrock Trail atop the Rims near the airport is informative and sobering.

The Eagle Sandstone that forms the Rimrocks was once a continuous layer deposited along the sandy shoreline of the Western Interior Seaway about 80 million years ago in Cretaceous time. The quartz-rich sandstone appears to be the remains of a beach on a barrier island, perhaps like Galveston Island in Texas, that once stood between a coastal lagoon west of Billings and the shallow inland sea that flooded Montana from the east. Waves and currents shifted sandbars around, leaving behind cross-bedding, while burrowing animals left traces of their activities. The sandstone was deposited at a time of gradual sea-level drop, and shoreline sand migrated eastward over fine-grained marine sand and mud of the underlying Telegraph Creek Formation. The rockfall hazard exists because the harder, more resistant sandstone is undermined as the softer shale erodes.

For tens of thousands of years, the Yellowstone River has been moving back and forth across its floodplain, eroding down through the rising plains. The horizontal layers of sandstone and shale were swept away in the river valley but persist as resistant rims at the valley's edge. The well-cemented sandstone resists erosion, forming a 500-foot-high step in the plains that slowly erodes through rockfall. As the underlying Telegraph Creek shale erodes back into the cliff wall, the overlying Eagle Sandstone is left overhanging and unsupported. Giant slabs of rock separate from the cliff face along vertical cracks in the sandstone, aided by moisture that seeps in and pries the rock apart when it freezes and expands. It is astonishing to see the giant cracks in the rock above the houses. Some rocks have detached from the

From I-90 in Billings, travel 4.1 miles northwest on MT 3 to the West Rims Overlook Parking Area. Walk the Rimrock Trail in either direction to see rockfall hazards. Continue 2.9 miles west to Zimmerman Park for more trails and views of the Yellowstone River Valley from the Rims.

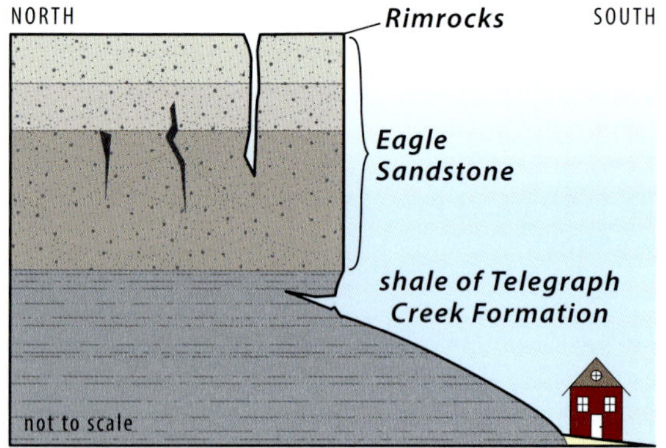

Sketch showing how erosion of the soft Telegraph Creek Formation shale undercuts the overlying Eagle Sandstone allowing it to fracture and fall. —Modified from US Geological Survey

A walk along the Rimrock Trail from the West Rims Overlook Parking Area provides a sobering view of past rockfalls from the Rims.

cliff and dropped several feet, while others have holes and arches eroded in the sandstone through which you can view the homes below.

It is imperative to find ways to mitigate the rockfall hazard because numerous structures already exist below the Rims. Montana Bureau of Mines and Geology maps identify those areas of highest risk for rockfall. Some stabilization techniques are being used to secure the slope. Rock bolts can be used to secure slabs of cracked sandstone, while soil nails and biowalls (retaining structures using plants) help stabilize the softer rock and soil below. Surface steel mesh is used to prevent loose rock from bounding uncontrollably down the slope, and rock or concrete retaining walls can also slow or stop falling boulders. Future construction should simply be avoided below the Rims.

Big cracks in the sandstone Rimrocks above Billings serve as a stark reminder of the significant rockfall hazards faced by those living below the Rims.

References Cited

Aram, R., 1979, Cenozoic geomorphic history relating to Lewis and Clark Caverns, Montana: Montana State University, M.S. thesis.

Azevedo, P.A., 1993, Tectono-sedimentary evolution of a Late Cretaceous alluvial fan, Beaverhead Group, southwestern Montana: Montana State University, M.S. thesis.

Boudreau, A.E., Butak, K.C., Geraghty, E.P., Holick, P.A., and Koski, M.S., 2020, Mineral Deposits of the Stillwater Complex, Montana Bureau of Mines and Geology (MBMG) Special Publication 122: Geology of Montana, v. 1, https://doi.org/10.59691/UFCN1290.

Christiansen, R.L., 2001, The Quaternary and Pliocene Yellowstone Plateau volcanic field of Wyoming, Idaho, and Montana: USGS Professional Paper 729-G, https://doi.org/10.3133/pp729G.

Foster, D.A., Mueller, P.A., Mogk, D.W., Wooden, J.L., and Vogl, J.J., 2006, Proterozoic evolution of the western margin of the Wyoming Craton: Implications for the tectonic and magmatic evolution of the northern Rocky Mountains: Canadian Journal of Earth Sciences, v. 43, p. 1601–1619, https://doi.org/10.1139/e06-052.

Fritz, W.H., and Thomas, R.C., 2011, Roadside Geology of Yellowstone Country, 2nd Edition: Missoula, Montana, Mountain Press Publishing.

Fullerton, D.S., Colton, R.B., and Bush, C.A., eds., 2012, Quaternary geologic map of the Havre 1° x 2° quadrangle, United States: Open File Report 2012–1028, scale 1:250,000.

Gammons, C.H., and Duaime, T.E., 2020, The Berkeley Pit and surrounding mine waters of Butte: MBMG Special Publication 122: Geology of Montana, v. 1.

Gammons, C.H., Metesh, J.J., and Duaime, T.E., 2006, An overview of the mining history and geology of Butte, Montana: Mine Water and the Environment, v. 25, no. 2, p. 70–75, https://doi.org/10.1007/s10230-006-0113-7.

Gammons, C.H., Korzeb, S.L., and Hargrave, P.A., 2020, Metallic ore deposits of Montana: MBMG Special Publication 122: Geology of Montana, v. 1.

Harlan, S.S., Snee, L.W., Reynolds, and others, 2005, $^{40}Ar/^{39}Ar$ and K-Ar Geochronology and tectonic significance of the upper Cretaceous Adel Mountains Volcanics and spatially associated Tertiary igneous rocks, northwestern Montana: USGS Professional Paper 1696, https://doi.org/10.3133/pp1696.

Harms, T.A., and Baldwin, J.A., 2020, Paleoproterozoic geology of Montana, MBMG Special Publication 122: Geology of Montana, v. 1.

Harms, T.A., Brady, J.B., and Cheney, J.T., 2006, Exploring the Big Sky Orogen in Southwest Montana: 19th Annual Keck Symposium, p. 171–176, https://keckgeology.org/wp-content/uploads/2024/07/montanaoverview-2006.pdf.

Hauptvogel, D., Sisson, V., and Comas, M., 2024, Investigating the Earth: Exercises for Physical Geology: Houston, Texas, UH Libraries.

Hearn, B.C., Jr., 1989, Introduction: T346 Montana High-Potassium Igneous Province, *in* Hearn, B.C., Jr., Dudas, F.O., Eggler, D.H., and others, eds., Montana high-potassium igneous province: Crazy Mountains to Jordan, Montana: American Geophysical Union, Field Trip Guidebooks, V. 346, p. 1–5, https://doi.org/10.1029/FT346p0001.

Horner, J.R., and Hanson, D.A., 2020, Vertebrate paleontology of Montana: MBMG Special Publication 122: Geology of Montana, v. 1.

Hyndman, D.W., and Thomas, R.C., 2020, Roadside Geology of Montana, 2nd Edition: Missoula, Montana, Mountain Press Publishing.

Keefer, W.R., 1971, The geologic story of Yellowstone National Park: USGS Bulletin 1347, https://doi.org/10.3133/b1347.

Lageson, D.R., Kalakay, T.J., and Foster, D.A., 2020, Mountain building: The orogenic evolution of Montana: MBMG Special Publication 122: Geology of Montana, v. 1.

Lonn, J.D., Burmester, R.F., Lewis, R.S., and McFaddan, M.D., 2020, The Mesoproterozoic Belt Supergroup: MBMG Special Publication 122: Geology of Montana, v. 1.

Losh, S., 2025, Block sliding, Heart Mountain Detachment, Wyoming, https://faculty.mnsu.edu/stevenlosh/research/block-sliding-heart-mountain-detachment-wyoming/

Lowenstern, J.B., and Hurwitz, S., 2008, Monitoring a supervolcano in repose: Heat and volatile flux and the Yellowstone Caldera: Elements, v. 4, p. 35–40, https://doi.org/10.2113/GSELEMENTS.4.1.35.

McCafferty, A.E., 1995, Assessing the presence of a buried meteor impact crater using geophysical data, south-central Idaho: Colorado School of Mines, M.S. Thesis.

McDonald, C., Elliott, C.G., Vuke, S.M., Lonn, L.D., and Berg, R.B., 2012, Geologic map of the Butte south 30′ x 60′ quadrangle, southwestern Montana: MBMG Open File Report 622.

Mogk, D.W., Mueller, P.A., and Henry, D.J., 2020, The Archean geology of Montana: MBMG Special Publication 122: Geology of Montana, v. 1.

Mosolf, J.G., 2023, A Hawaiian-style lava flow in southwestern Montana: Yellowstone Caldera Chronicles, May 1, Yellowstone Volcano Observatory.

Mosolf, J.G., Kalakay, T., and Salazar, R., 2025, Field guide to Bannack State Park, southwest Montana, *in* Elliott, C., and A. English, eds., Bannack, Montana: Northwest Geology, v. 54, p. 141–157.

Morgan, L.A., Pierce, K.L., and Shanks, W.C.P., 2008, Track of the Yellowstone hotspot: Young and ongoing geologic processes from the Snake River Plain to the Yellowstone Plateau and Tetons, *in* Raynolds, R.G., ed., Roaming the Rocky Mountains and Environs: Geological Field Trips: Geological Society of America Field Guide 10, p. 139–173, https://doi.org/10.1130/2007.fld010(08).

Parker, S.D., and Pearson, D.M., 2021, Pre-thrusting stratigraphic control on the transition from a thin-skinned to thick-skinned structural style: An example from the double-decker Idaho–Montana fold–thrust belt: Tectonics, v. 40, doi: https://doi. org /10.1029/2020TC006429.

Reed, M., and Dilles, J., 2020 Ore Deposits of Butte, Montana: MBMG Special Publication 122: Geology of Montana, v. 1.

Sears, J.W., Hendrix, M.S., Thomas, R.C., and Fritz, W.J., 2009, Stratigraphic record of the Yellowstone hotspot track, Neogene Sixmile Creek Formation grabens, southwest Montana: Journal of Volcanology and Geothermal Research, v. 188, no. 1, p. 250–259, https://doi.org/10.1016/j.jvolgeores.2009.08.017.

Scarberry, K.C., Yakovlev, P.V., and Schwartz, T.M., 2020, Mesozoic magmatism in Montana: MBMG Special Publication 122: Geology of Montana, v. 1.

Smith. L.N., Hill, C.L., and Reiten, J., 2020, Quaternary and late Tertiary of Montana: Climate, glaciation, stratigraphy, and vertebrate fossils: MBMG Special Publication 122: Geology of Montana, v. 1.

Stickney, M.C., 2020, Earthquakes and seismographic monitoring in Montana: MBMG Special Publication 122: Geology of Montana, v. 2.

Stickney, M.C., Haller, K.M., and Machette, M.N., 2000, Quaternary faults and seismicity in western Montana: MBMG Special Publication 114, scale 1:750,000.

Thomas, R.C., and Sears, J.W., 2020, Middle Miocene through Pliocene sedimentation and tectonics in Montana: A record of the outbreak and passage of the Yellowstone hotspot: MBMG Special Publication 122: Geology of Montana, v. 1.

Torsvik, T.H., Gaina, C., and Redfield, T.F., 2008, Antarctica and global paleogeography: From Rodinia through Gondwanaland and Pangea to the birth of the Southern Ocean and the opening of gateways, *in* Cooper, A.K., Barrett, P.J., Stagg, H., Storey, B., Stump, E., and Wise, W., eds., Antarctica: A Keystone in a Changing World: Proceedings of the 10th International Symposium on Antarctic Earth Sciences: Washington DC, The National Academies Press, p. 125–140, https://doi.org/10.17226/12168.

Vuke, S.M., 2020, Synorogenic basin deposits and associated Laramide uplifts in the Montana part of the Cordilleran foreland basin system: MBMG Special Publication 122: Geology of Montana, v. 1.

Vuke, S.M., 2020, The Eocene through early Miocene sedimentary record in Western Montana: MBMG Special Publication 122: Geology of Montana, v. 1.

Vuke, S.M., and Colton, R.B., 2003, Geologic map of the Terry 30′ x 60′ quadrangle, eastern Montana: MBMG Open-File Report 477, scale 1:100,000.

Winston, D.W., 1989, Introduction to the Belt, in Volcanism and plutonism of western North America, v. 2, field trip T334, Middle Proterozoic Belt Supergroup, western Montana, trip leaders D. Winston, R. J. Horodyski, and J. W. Whipple, in the collection of field trips for the 28th International Geological Congress, American Geophysical Union, 1–6.

Yakovlev, P.V., Scarberry, K.C., and Stickney, M.C., 2020, Neotectonic development of Western Montana: MBMG Special Publication 122: Geology of Montana, v. 1.

Glossary

alkalic. Said of igneous rocks that contain an uncommonly large amount of sodium, potassium, or both.

alluvial fan. A fan-shaped deposit of sediment produced where a stream flows from a mountain onto a valley floor. Mudflows can also deposit them.

ammonite. A coiled-shell marine animal, related to octopus, that lived from Devonian to Cretaceous time.

andesite. A volcanic rock intermediate between rhyolite and basalt in composition, color, and eruptive behavior.

anticline. An arch folded in layered rock.

asthenosphere. The mechanically weak (ductile or "plastic") part of Earth's upper mantle that lies below the more rigid lithosphere and above the deeper mantle.

badlands. A ragged, deeply gullied landscape of dry, soft sedimentary rock severely eroded by wind and rain; characterized by hoodoos and mushroom-shaped rocks.

basalt. A common volcanic igneous rock. Basalt is black and consists mostly of the minerals plagioclase and pyroxene.

basement rocks. A general term referring to ancient, strongly metamorphosed rocks that underlie at deeper levels than most rocks exposed at the surface. In Montana, they are older than the Belt Supergroup and commonly occur as schist, gneiss, and granite.

Basin and Range. A region of wide, north-trending valleys and narrow ranges centered on Utah and Nevada and extending north into Montana east of the Idaho batholith but west of the high Rockies. It formed by stretching of the continental crust.

batholith. Any large mass of granite or similar rock, technically one that covers an area greater than about 40 square miles

bedding. The layering in sedimentary rocks that formed during sediment deposition because of variations in sediment size and composition with time.

bedrock. The solid rock of the Earth's crust. It often underlies loose sediment or soil.

biotite. The black to dark-brown, potassium-magnesium-iron variety of mica common in granites and light-colored gneisses and schists.

brachiopod. A clam-like marine animal that appeared in Cambrian time and continues to exist today. It can be distinguished from a clam by its shell symmetry, which is perpendicular to the hinge line.

breccia. A rock composed of angular fragments.

calcite. A mineral composed of calcium carbonate. It is a major component of limestone.

caldera. A large, collapsed depression that sank into the magma chamber after a volcano erupted explosively.

carbonate. Rocks such as limestone or dolostone that are formed with calcium carbonate or magnesium carbonate.

chert. A sedimentary rock composed of quartz crystals too small to see with the naked eye. It forms through the accumulation of silica shells of tiny plants and animals, by precipitation directly from water, or as a chemical deposit precipitated from groundwater percolating through limy sediments after deposition.

coarse-grained. Said of sedimentary rocks that have particles that are relatively large, usually averaging 2 millimeters in diameter or larger. Also said of igneous rocks with relatively large crystals.

compressional fault. *See* reverse fault.

conglomerate. A sedimentary rock consisting of rounded pebbles surrounded by finer grains.

contact metamorphism. Temperature-dominated metamorphism caused by intrusion of hot magma.

crinoid. A shallow marine plant growing from the seafloor on a thin stem.

cross-beds. Internal sedimentary layers more steeply oriented than and at an angle to the main layers. Produced by the movement (bouncing) of grains carried in water or air currents.

crust. The Earth's outer layer. It is about 4 miles thick under the oceans and tens of miles thick under the continents.

crystalline rock. Grainy igneous or metamorphic rock such as granite, schist, or gneiss.

cyanobacteria. Bacteria that obtain their energy through photosynthesis and produce oxygen as a result. They first appeared in Proterozoic time and built layered sedimentary structures called stromatolites.

debris flow. A mass flow of water, mud, and rocks.

detachment. A nearly horizontal or gently dipping surface or fault associated with large-scale extensional tectonics and often with large displacement (tens of miles) that juxtapose lightly metamorphosed rocks above from higher-grade metamorphic rocks below.

diabase. An igneous rock that resembles basalt except that individual crystals of black augite and pale plagioclase feldspar are visible without using a microscope. Grain size is intermediate between gabbro and basalt.

dike. A body of igneous rock that formed as a sheet-like body of magma cut steeply across existing rock.

dip. Slope of a geologic surface, such as a fault or bedding plane, measured down from the horizontal.

dolomite. A sedimentary carbonate rock composed of the calcium-magnesium carbonate mineral dolomite.

drumlin. A streamlined hill composed of till laid down beneath a moving glacier.

dunes. Fine-grained sand grains accumulated in piles by bouncing in the wind.

erosion. The movement or transport of weathered material by water, ice, wind, or gravity.

extension. Stretching or pulling apart of Earth's crust.

extensional fault. A fault, generally with a high angle, that results from the extension of the crust. Also called a *normal fault*.

fault. A fracture in Earth's crust along which the opposite sides slipped past each other.

fault scarp. The steep slope produced by slip on a fault surface.

feldspar. A group of abundant aluminum silicate minerals. Potassium feldspar, called orthoclase, is typically white, beige, or pink. Sodium-calcium feldspar, called plagioclase, is typically white or greenish white. Most feldspars occur in rectangular crystals.

fine-grained. Said of sedimentary rocks that have particles that are relatively small, usually averaging less than 2 millimeters in diameter. Also said of igneous rocks with relatively small, hard-to-see crystals.

fold-and-thrust belt. A long zone of deformation of layered sedimentary rocks crumpled into folds and broken by thrust faults.

foreland basin. A big, elongate depression formed as Earth's crust is depressed at the margin of a newly formed mountain range.

formation. A formal stratigraphic unit of sedimentary rocks having similar characteristics such as rock composition.

fossil. Any trace of a plant or animal preserved in rock.

gabbro. A coarse-grained igneous rock composed principally of the minerals plagioclase feldspar and pyroxene. Gabbro is a coarse-grained equivalent of basalt.

gneiss. Coarse-grained metamorphic rocks having a banded or streaky appearance. Gneisses form through recrystallization at high temperature and pressure of various kinds of sedimentary or igneous rocks.

graded beds. Sedimentary layers in which the grain size is coarser at the bottom and becomes finer toward the top.

granite. A coarse-grained igneous rock composed of more than 35 percent alkali feldspar and more than 20 percent quartz. An igneous rock is said to be **granitic** if it has coarse-grained, light-colored crystals.

granodiorite. A rock similar to granite but with less alkali feldspar and typically more dark minerals like biotite and hornblende.

groundwater. Water in the ground, mostly derived from rainfall and snowmelt.

group. A formal rock unit containing two or more formations.

hornblende. A common silicate mineral that crystallizes into glossy black needles in light-colored igneous and metamorphic rocks such as granite, schist, and gneiss.

hot spot. Site of a large volcano erupted above a rising plume of heat from deep in the mantle.

ice cap. The large expanse of ice over a high mountain range.

igneous. Rocks that crystallized from magma or lava.

intrusion. Igneous rocks that have intruded into and cooled within older rocks.

karst. Weathering of limestone to form cavities and sharply ragged surfaces.

laccolith. An igneous intrusion that formed when a blister of magma injected between layers of sedimentary rock.

Laramide deformation. The rise of thick, basement-cored crustal blocks on steep reverse faults from Late Cretaceous to earliest Cenozoic time.

lava. When molten rock, or magma, reaches Earth's surface and erupts, it is called lava.

limestone. A sedimentary rock composed of the mineral calcite, a calcium carbonate.

lithosphere. The rigid, outer rind of the Earth, approximately 60 miles thick, including the crust and uppermost mantle.

magma. Molten rock beneath Earth's surface; magma becomes lava if it pours out on the surface.

magma chamber. A large mass of molten magma within Earth's crust that may feed a volcano.

mantle. A deep internal layer, below Earth's crust, that makes up most of the volume of the Earth.

marble. A grainy metamorphic rock formed by the recrystallization of limestone or dolomite.

marine. Related to the sea or ocean.

meander. The sinuous, back and forth sweep of a river or stream.

member. A formal stratigraphic unit representing some specially developed part of a formation.

metamorphism. Recrystallization of a rock at high temperature, pressure, or both to produce a new rock that differs considerably in texture from the original and has different minerals but may still have a similar chemical composition.

mica. A group of silicate minerals that tend to split into thin flakes. The common varieties are black biotite and colorless muscovite. Both typically occur in generally light-colored igneous and metamorphic rocks such as granite, gneiss, or schist.

moraine. A deposit of glacial till left at the fringe of a melting glacier.

mylonite. A generally fine-grained, intensely sheared rock in a fault zone.

normal fault. A fault along which one fault block slides down a sloping surface relative to the fault block on the other side of the fault. A normal fault forms as the result of forces that are pulling or stretching an area apart.

North American plate. The major lithospheric tectonic plate that includes North America and the western half of the Atlantic Ocean.

offset. Used to describe the distance of separation along a fault.

olivine. A glassy green, iron, and magnesium silicate mineral, common in basalt and gabbro.

ore. Any kind of rock that can be profitably mined.

outwash. Sediment deposited from glacial meltwater. Like other stream sediments, outwash typically consists of distinct layers of well-sorted sand, gravel, and occasionally clay.

peat. An accumulation of partially decayed plant material. Buried peat may eventually turn into coal over time.

pediment. A broad, very gently dipping erosional surface, often at the mouth of one or more mountain canyons and sometimes covered by a layer of gravel.

pegmatite. An extremely coarse-grained igneous rock, most commonly of granite composition and formed in the later stages of crystallization of granite magma.

plagioclase. An extremely common feldspar mineral that contains calcium and sodium in varying proportion. It is typically white, greenish white, or gray and occurs in a wide variety of igneous and metamorphic rocks.

plate. One of several large segments of the lithosphere, the rigid upper part of Earth's crust and upper mantle.

pluton. A solidified body of intrusive igneous rock, such as granite, that crystallized beneath the surface. A cluster of plutons may be large enough to form a granite batholith.

pumice. A light-colored, frothy, sponge-like volcanic rock blown out of volcano from a silica-rich magma containing expanding gases.

pyroclastic flow. A dense mixture of volcanic ash, fragments, and gas that rushes, 100 or more miles per hour, down the flank of a volcano. These flows collapse from tall columns of erupted material.

pyroxene. A black or dark green silicate mineral that typically composes about half of basalt or gabbro. It is rich in calcium and magnesium, with some iron and aluminum. Augite is a common variety.

quartz. A very common mineral composed of silicon dioxide. Quartz comes in many varieties of which the most common is clear and colorless, like glass.

quartzite. A sandstone metamorphosed by heat, pressure, or both so that the rock breaks across individual sand grains rather than around them.

resistant. Said of a rock or rock outcrop that withstands the effects of weathering or of erosion.

reverse fault. A fault in which the block overlying a dipping fault appears to have moved up relative to the block underlying the fault. Reverse faults commonly form because of compression.

rhyolite. A pale volcanic rock, either ash or lava, that contains quartz. Rhyolite has about the same composition as granite.

rift. A long, narrow, down-dropped zone where Earth's crust is separating. The process of a tectonic plate being pulled apart or splitting is called **rifting**.

rimrocks. The sheer cliffs at the edge of a plateau or narrow canyon.

sandstone. A deposit of sand hardened into rock by cementation.

scarp. A linear cliff produced by faulting or erosion. The term is an abbreviation for escarpment.

schist. A metamorphic rock that contains enough parallel mica flakes to make it split into sheets or enough needle-shaped minerals to make it splintery. Schists form through recrystallization at high temperature and pressure of various kinds of sedimentary rocks.

sedimentary rocks. Deposited material such as limestone, sand, clay, mud, or gravel hardened into rock.

Sevier deformation. Thrust faulting and folding of sedimentary rocks primarily during Cretaceous and earliest Cenozoic time.

shear. The action of two bodies of rock sliding past each other. The sheared part of the rock is called a **shear zone**.

shonkinite. A dark igneous rock composed mostly of augite (a pyroxene mineral) and potassium feldspar. It is similar in composition to basalt, except in having a much larger potassium content.

silica. Silicon dioxide, the compound that makes up quartz in all its varieties, including chert.

sill. A sheet-like body of igneous rock sandwiched between layers of sedimentary rocks.

shale. A soft, flaky rock composed largely of clay.

skarn. A kind of metamorphic rock that forms where granitic magma reacts with limestone. Skarns typically consist of large crystals of garnet, epidote, and other calcium silicate minerals.

smelter. A large mill in which a metallic mineral concentrate is roasted in the presence of a reducing (low-oxygen) agent to separate molten metal from sulfur as sulfur-oxide gases.

stock. A granite intrusion or pluton smaller than 40 square miles in area. A cluster of stocks may be large enough to form a granite batholith.

stratovolcano. A large volcano, usually andesite to basalt in composition, consisting of layers of volcanic ash and lavas.

subduction zone. A major tectonic collision zone, most commonly between a descending oceanic plate and an overriding continental plate. It can cause major earthquakes and produce explosive volcanoes in the upper plate.

sulfide. A mineral compound of a metallic element and sulfur, for example pyrite (iron sulfide) or chalcocite (copper sulfide).

supercontinent. An amalgamation of continents; for example, Rodinia and Pangea.

syncline. A downward bend or trough folded in layered rocks.

tailings. Waste rock produced during mining.

tectonics. The study of regional-scale deformation of Earth's crust caused by interactions between tectonic plates.

tephra. Volcanic ash and other fragmental material blown from a volcano.

terrace. Older stream deposits standing at a somewhat higher level than the active stream deposits.

thrust fault. A gently dipping (less than 45 degrees) fracture along which the rocks above the fault surface moved up and over those beneath. The rock above a thrust fault is termed a **thrust sheet,** which is thin relative to its areal extent.

till. Sediment dumped directly from glacial ice. Till typically consists of partly rounded rocks of all sizes and shapes and without internal layering.

trilobite. An extinct group of arthropods, with external shells, that appeared in Cambrian time about 520 million years ago and were exterminated in the global extinction at the end of Permian time, 252 million years ago.

tuff. A soft volcanic rock composed of volcanic ash, most commonly rhyolite.

unconformity. A gap in the rock record, either because sediments were not deposited during that time or because they eroded before the deposition of overlying rocks.

vein. A fracture filled usually with quartz or calcite but sometimes containing more valuable minerals.

weathering. The process by which rocks break down near Earth's surface due to exposure to air, water, and the action of organisms.

Index

About the Author

Robert C. Thomas is a professor of geology in the environmental sciences department at the University of Montana Western in Dillon. He earned a BA from Humboldt State University, an MS from the University of Montana, and a PhD from the University of Washington. A Montana Regents Professor and Teaching Scholar, Montana Educator of the Year, and a Carnegie US Professor, Rob is a Fellow of the Geological Society of America and recipient of their Distinguished Service Award. He has authored or coauthored more than 75 publications, including *Roadside Geology of Montana* and *Roadside Geology of Yellowstone Country*. As a field geoscientist, he engages students in real projects designed to benefit society and the environment. His personal passions are family, mountain recreation, and playing guitar around a campfire.

About the Illustrator

Chelsea M. Feeney spent her youth in the foothills of the Blue Ridge Mountains in central Virginia and earned her BS in geosciences from Virginia Tech. Pursuing her interest in landscape evolution, she headed west and received her MS in geology from the University of Montana, publishing her first map through the Montana Bureau of Mines and Geology. She worked for several organizations in her early career, including Virginia's Division of Geology and Mineral Resources, Oregon's Department of Transportation, the Wild Rockies Field Institute, and Yellowstone National Park. In 2009, she began working with Mountain Press Publishing Company producing maps and figures for their renowned Roadside Geology guidebook series, now published through the Geological Society of America. Chelsea runs her own custom cartographics business, and her work is featured in more than 100 publications, books, and journal articles (www.cmcfeeney.com). Chelsea lives in Moscow, Idaho, with her husband, a fellow geologist, and their two children.